What's Your PROBLEM?

Identifying and Solving the Five Types of Process Problems

What's Your PROBLEM?

Identifying and Solving the Five Types of Process Problems

Kicab Castañeda-Méndez

CRC Press is an imprint of the
Taylor & Francis Group, an **informa** business
A PRODUCTIVITY PRESS BOOK

CRC Press
Taylor & Francis Group
6000 Broken Sound Parkway NW, Suite 300
Boca Raton, FL 33487-2742

Printed in the United States of America on acid-free paper
Version Date: 20120627

International Standard Book Number: 978-1-4665-5269-2 (Paperback)

Library of Congress Cataloging-in-Publication Data

Castañeda-Méndez, Kicab.
What's your problem? : identifying and solving the five types of process problems / Kicab Castañeda-Méndez.
p. cm.
Includes bibliographical references and index.
ISBN 978-1-4665-5269-2
1. Problem solving. 2. Process control. 3. Reengineering (Management) I. Title.

HD30.29.C373 2013
658.4'03--dc23 2012024327

Visit the Taylor & Francis Web site at
http://www.taylorandfrancis.com

and the CRC Press Web site at
http://www.crcpress.com

To my mother,

who taught me to keep it simple.

Contents

Preface

You want to solve process problems and get rid of defects. It doesn't have to be difficult—apply the Pareto principle.

This book will show you how to teach your employees to solve not all process problems, but the *most common*. And, it will show you how to do it more effectively and much quicker.

The major premise of this book is that there are only five types of process problems and, therefore, the Six Sigma methodology (define, measure, analyze, improve, control (DMAIC)) can be vastly simplified for learning, applying, teaching, and mentoring. In addition, the approach in this book discusses how to seamlessly integrate Lean and Six Sigma methodologies and their purposes.

Two types of problems virtually occur in every process. A third type of process problem is frequent in manufacturing and a fourth occurs in forecasting processes everywhere in the company, not just in finance. The fifth type of process problem is commonly addressed incorrectly. The correct way to address this problem will help employees and, eventually, the company to become more customer-focused.

What's Your Problem? Identifying and Solving the Five Types of Process Problems shows you how you can teach yourself, apply what you have learned on your own, coach others, and develop short and effective courses for instructing personnel.

After successfully improving processes and solving problems for decades under the guise of statistical support, quality circles, TQM (total quality management), re-engineering, cycle-time reduction, Lean, kaizen, Deming's 14 points, Juran's Trilogy, Baldrige and state quality awards, and Six Sigma, I have learned one thing that is relevant and is the reason for this book.

Practice what you preach and teach—This book is meant to instruct you on how to reduce learning time by 50% to 80%, if self-learning and, for others, if creating training courses. In addition, if you are coaching others, the methodology you will learn will help reduce the time in completing projects, reduce documentation of projects, and increase understanding of the project on which you are working. You also can improve your courses (participants will learn faster, acquire essential skills sooner, and retain more) and at a much lower cost.

Introduction

Philosophy on Problem Solving

Solve specific problems with specific solutions. Learn to recognize different types of process problems to solve them more effectively and efficiently.

Use Lean principles.

Don't pass on defects. Learn how to solve the problem of a single defect before moving to solving chronic problems. This fosters a culture of problem solving and continuous improvement.

Process

A process is a set of actions resulting in an output whose purpose is to repeatedly satisfy specified requirements.

Repetitiveness allows you to improve process performance. One-time events cannot be improved. Since much of what a business (and a person in his/her life) does is repetitive, process improvement is applicable. It should be a company's ongoing strategy and a skill of every employee and individual.

Process Performance

Because of the repetitiveness, to improve process performance means to do better on recent repetitions compared to past repetitions. That is, to improve means to do better one time period (with one or more repetitions) compared to a previous time period (with one or more repetitions).

Satisfying the specified requirements is good, so process performance can be defined as:

The performance of a process is the percent good for a specified period.

Process Problems

A problem is a gap. You can think of a gap as the difference between what you have and what you want or need. Process problems are defects. When a process output does not meet requirements, it is a defect. If you don't want this defect, then you have a gap. Process performance is the frequency of outputs meeting requirements. When this frequency is not at the desired level, you have another gap.

Generally defects occur when:

- We didn't do something that was needed or we did something that wasn't needed.
- We did something that was needed but did it wrong.
- We did something that was needed but not the best way it could be done.

Recognizing specifically how these different cases occur helps you solve problems faster. How a defect occurs is the cause. Different causes typically result in different types of defects.

There are five types of process problems based on the type of cause. Probably, you have already solved these types of problems.

1. Do you use auto-correct or word suggestion in texting or emailing or writing documents? Then you have reduced the number of defects due to errors.
2. Do you have specific places for each type of clothes you own (e.g., underwear, socks, tops, pants) or have you used a short cut? Then you have prevented defects due to delays.
3. Have you ever adjusted a thermostat to be more comfortable? Then you have solved a defect due to suboptimality.
4. Do you know when you have to leave home to get to work on time or adjusted the amount of perishable food you buy to reduce waste? Then you have solved a defect due to unpredictability.
5. Have you ever asked a friend (partner, parent, child) why they did something you objected to and resolved the issue when you got the explanation? Then you have solved a defect due to personal reason.

This book will show you how to take advantage of this knowledge, skills, and experience. This will reduce variation in teaching and learning (adopting Six Sigma principles for defect reduction), reduce class time (adopting Lean principles for defect reduction), and increase learning (meeting your requirements for efficient and effective learning).

Solving Process Problems

Often we focus not on the process, but the resulting defect. Of course, we want to correct the defect, but we also want to prevent that defect from reoccurring. Defects are the effects of causes. You need to identify the specific causes. Solving process problems means reducing the probability of that defect occurring again. To do this, you have to take action.

There are two types of actions: corrective and preventive. Suppose you order an item and it doesn't arrive. You inform the company and the company then sends you the item or it sends a coupon for a discount on your next purchase. These are corrective actions. The company only corrected the defect or placated you, the customer. It did not change the probability of the same defect occurring again.

Corrective action eliminates the defect through rework or replacement. This eliminates the problem as a gap between the defective output unit you received and you not wanting that defective unit.

Another gap was between getting a defect and not wanting defects in the future. Preventive action reduces (or eliminates) the probability of the defect reoccurring.

When the company changes the process to reduce the frequency of defects, then it has taken preventive action. This does not address *your* defect, but reduces the probability of future defects from this error occurring.

Ideally you want to take both corrective and preventive actions. Process improvement is preventive action. It solves process problems and reduces their probability of reoccurring. This book is about preventive actions.

Continuous Improvement

Rather than take a project approach, you should consider the kaizen approach in its original sense as incremental improvement. This requires changing your thinking and behavior from "once-and-done" to "everything-can-always-be-improved" and "what-can-I-improve-today" thinking. The latter thinking and behavior result in continuous improvement.

This book can help you create the continuous improvement thinking and behavior. It practices what it preaches; it simplifies process problem solving. If it is complicated, then you are not going to do it regularly. If you don't do it regularly, you won't change your thinking and behavior. The simpler it is, the easier it is to become a habit. It has to be a habit in order for your thinking and behavior to change.

There are three sections in this book. Section I (Single Unit Process Improvement) starts your journey to fast and effective problem solving by making improvements as soon as you find a single defect. In this section, you learn how to close the gap between having a defect and not wanting this defect. Because you are working on a single defect, you will learn a simple three-step procedure for addressing this defect no matter which of the five problem types it is.

The first two chapters are enough to help you begin the transformation. In just an hour or so, you should be able to understand the two process problems error and delay thoroughly enough to begin your journey. Then you can begin practicing immediately on whatever process you want.

Chapters 3 and 4 briefly describe the two process problems of suboptimality and unpredictability. Statistics and software are often required to solve these problems.

Chapter 5 presents a unique type of problem. At first blush, it may seem so esoteric that you may ask why include it. However, I think you will find that personal reasons occur often enough (you may very well be "guilty" of using this approach to cause defects) that it is worthwhile to be aware of it more formally.

Section II (Multiple Unit Process Improvement) expands on the three-step procedure. When you have chronic process problems, it helps to be more critical in your thinking. The three-step procedure is expanded by one step. In addition, to make it clearer why you are doing these steps, each step is based on a question. The first four chapters of Section II show you what information you need and where you can get it to answer each of these four critical thinking questions.

Chapter 10 then shows you how to apply the same critical thinking questions to designing processes with better performance.

Learning Theories

Recent studies show that superior pattern recognition ability is critical to achieving superior performance. Practice alone is not enough.

General Problems

Gap between actual and desired

Process Problems

(Process Purpose: Produce outputs that repetitively meet requirements)
Output Defect = Requirement not met

Process Problem Situations

Single Output Problem: one output has defect
Multiple Output Problem: % defect too high

Process Problem Types by Cause

- Delay
- Error
- Suboptimality
- Unpredictability
- Personal Reason

Figure I.1 Problem hierarchy.

Many repetitions on the right things are the key to acquire a deeper understanding and instinctual habit of recognizing patterns.

This book takes this approach from the beginning. It starts with the premise that a specific type of problem is easier to solve than a general one. Section I describes five patterns rather than lumping all process problems together. Section II builds on these patterns by providing more practice on more complex circumstances. Figure I.1 describes the problem hierarchy starting with a definition of a general problem, then generally defining process problems. This is followed by the two process problem situations: single or multiple output units. Each case is then solved specifically as one of the five types of process problems.

Section III (Teaching Problem Solving) has two chapters on teaching the methods of this book. Chapters 11 and 12 have 14 lessons that you can use to create courses that teach the material in this book in the manner in which the book is written.

These chapters describe how to create classes that will get people solving process problems and eliminating defects as soon as possible. These lessons combine practice, frequency, and simplicity as keys to changing thinking and behavior. It also includes suggested curricula.

Chapter 13 provides some guidance for those who want to mentor others on these methods. The mentoring guidelines also are based on repetition, pattern recognition of the five process problems, and simplicity.

Benefits

Solving process problems can be simplified to three steps:

1. Identify the type of problem.
2. Find the root cause.
3. Address the root cause.

Skipping the first step results in wasted time and risking not achieving the last two steps.

When we fail to distinguish the types of problems, then we fail to distinguish efficient ways of finding root causes and determining solutions. We don't know what tools to use, so we use all the ones we learned. We don't know what information we need, so we gather as much as possible. We don't know how to focus on plausible root causes, so we generate as many as we can think of.

Imagine a problem at home. If you didn't distinguish between it being an electrical problem or a plumbing problem or a leaky roof problem or bug infestation problem, then you would not know where to begin. Calling a plumber when the problem is the roof or termites would only waste time and befuddle the plumber.

Rather than looking everywhere for a root cause, knowing the type of defect can help you narrow the search. Data do not actually have to be collected to find root causes for error and delay problems.

Book Format

This book follows the format for teaching discussed previously in the sections on Continuous Improvement and Learning Theory. That format is based on the philosophy of repeated practice starting with the basics and adding complexity as mastery occurs.

Each section starts with a short introduction. The introduction explains the process improvement principle(s) that form the basis for the section. It also includes the learning format used in the section. The chapters of a section provide learning on the concepts and tools based on the principles of the section.

Let's begin.

Acknowledgments

First and foremost, I thank my wife, Marta, for her indefatigable support and encouragement. For this book in particular, she provided invaluable feedback on earlier drafts with respect to the writing, structure, and content. Conversations, discussions, but especially debates and counterviews with many people over several years led to the development and refinement of ideas presented in this book. While I can't recall the precise person with the precise help, I thank generally my colleagues and friends at or from GE Capital, Microsoft, Bausch & Lomb, and Merck. I also thank Michael Sinocchi, Jessica Vakili, Jay Margolis, and Maggie Mogck at Taylor & Francis Group for making this book possible and better.

SINGLE UNIT PROCESS IMPROVEMENT

Principle

Lean is founded on five principles:

1. Identify value from the end customer's view
2. Map the value stream
3. Create flow
4. Create pull
5. Seek perfection

In this section, you will use principle 3 with a simple understanding of principle 1. For this section, value will simply mean "defect free."

To create flow, you cannot stop the process. However, there is one exception—when there is a defect. Passing on a defect, not only does not add value, but the cost associated with that defect increases dramatically.

One way to create flow for value is to use the technique Sakichi Toyoda (founder of Toyoda Group in the late 1800s) developed. He created the first automated or self-powered loom. It had a mechanism that identified a particular defect: broken thread. The mechanism not only detected the defect but stopped the loom. This two-step process of detection and stoppage became known as *jidoka* or automation with a human touch.

To most effectively apply jidoka, you should process things one unit at a time.

Learning

In this section, you will learn to improve processes one unit at a time and one defect at a time. Because you are focusing on one unit and one defect, you will use a simple three-step procedure for addressing these problems:

1. Identify the type of problem by cause
2. Find the root cause
3. Address the root cause

Defects are not value added. Problems result in defects. Chapters 1 through 5 show how to use this three-step procedure to solve each type of problem. Each chapter presents a problem type in a simple format with five main sections:

- How to identify the type of defect (with key words to look for in problem descriptions and problem statements).
- How to find the root cause for that type of defect.
- How to address the root cause for that type of defect.
- One or more examples of applying the three steps.
- Suggestions of what you can do now to acquire skills and change your habits.

Chapter 1

Delay-Caused Defect

How to Identify Delay-Caused Defects

"What took you so long?"
"There was a huge line…"
"I made stops at…"
"I was put on hold…"
"They didn't have…"

This is probably the easiest type of process problem to identify. All process or cycle time issues are delay problems. In other words, whenever you want to reduce the time it takes to do something, it's a delay problem.

Key Words: on-time delivery, too long (e.g., lead time), too slow (e.g., pace, changeover, setup), backlog, (cycle, processing) time.

How to Find Root Causes of Delay-Caused Defects

Identify the type of delay occurring and where it occurs. Delays are root causes of an extended cycle time. Therefore, once you know where the delay is and its duration, you have identified not only the root cause, but its impact.

The key to successfully reducing cycle time through delay identification is to focus exclusively on the "thing" that is going through the process. For some processes, especially manufacturing ones, it is easy to identify the thing. For others it is not as clear.

To illustrate, in a hiring process, a candidate can be the "thing." Rather than looking at whether the Human Resources (HR) personnel and hiring manager are busy, look at whether the candidate is waiting, e.g., waiting for a response to a resume submission, to schedule an interview, to take the next step. It might be argued that a company has to consider several candidates and, therefore, the early candidates interviewed will have to wait until the others also have been interviewed. True. And, it is also true that the candidates will be waiting for the next step in the process.

There may be more than one "thing" that goes through the process. In the hiring process, the resume is also the "thing" and so is the evaluation of each candidate. If either is waiting (e.g., the resume or evaluation is in an inbox waiting for someone to act on it), then there is a delay.

Process time gets extended for these four reasons:

- Stopping the flow: Process is not acting on the thing.
- Repeating the flow for rework: Process is correcting defects on the thing.
- Doing nonvalue-added actions: Process is acting on the thing in a way that is valueless.
- Doing value-added actions slowly: Process is slowly acting on the thing in a way that is of value.

Each reason for delay must be identified as each has a different solution. In Figure 1.1, can you identify where the four reasons for delay occur?

TPS (Toyota Production System) or Lean is the original source of the tools and concepts of removing time from processes. Lean aims to remove

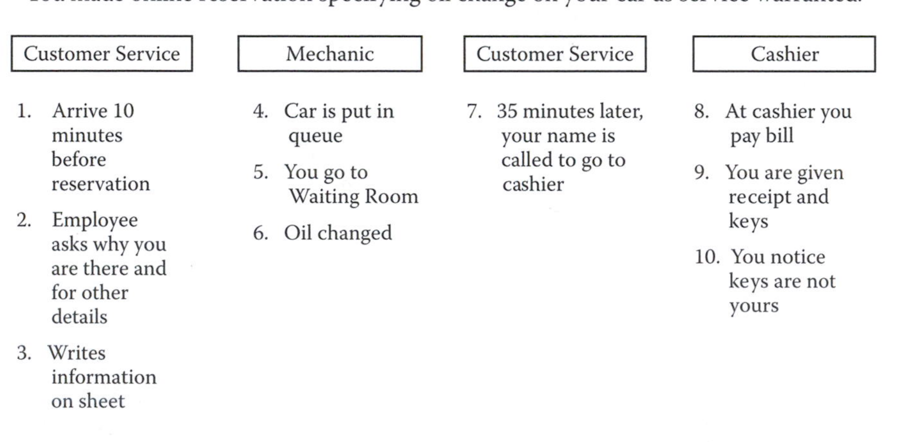

Figure 1.1 Reasons for Delays.

all waste throughout the value stream (i.e., process). It does this by using the five principles:

1. Define value
2. Create value stream
3. Create flow
4. Pull
5. Seek perfection

These principles are the basis for the four reasons for delays.

Delay Due to Flow Stopping

Flow stopping violates the third principle of creating flow. To see when the flow stops, track the "thing" that goes through the process and identify where no action is being taken on it. Note that the "thing" could be moving, but process activity is not the "thing." Flow of action has stopped because nothing is being done to it.

For example, an invoice is moving when it is carried from one location to another, but nothing is being done to it. It is waiting for something else to happen. A line moves giving the appearance of flow. However, notice that a moving line is no different than everyone seated waiting for the next number to be called.

Based on discussions among those who work the process, maps of the process are often drawn to identify stops.

You also can ask the people running the process where or when the "thing" is inactive.

Another less common, but often better, way to identify where the flow stops is to see the process. (This is called going to the gemba in Lean language.) People will forget things that they do when they do it often, routinely, or are busy. Someone might not recognize that the "thing" is waiting, say in an inbox, because they are busy doing other things.

Delay Due to Rework

Rework is a result of defects. Passing on defects violates the fifth principle of seeking perfection. Stopping because of defects is the only exception to creating flow.

Chapters 2 through 5 address defects that cause delays. Delays due to error-caused defects are addressed in Chapter 2 and delays due to other types of defects are addressed in Chapters 3 through 5. Learn how to address these other causes of defects.

Note that legitimately reducing rework is the only way to simultaneously address three factors in the right direction:

- Quality (less rework means higher quality)
- Cost (less rework cost)
- Time (less delay time)

Delay Due to Nonvalue-Added Actions

Creating nonvalue-added actions violates the first two principles. The first is defining value. Value-added actions are viewed from the customer's point of view, thus, they are more appropriately labeled customer value-added (CVA). CVA actions meet three criteria:

- Customer cares enough about it to pay for it.
- It physically changes the fit, form, or function.
- It is done right.

Fit refers to the thing's ability to interact or connect with other things with which it is intended to interact or connect. *Form* refers to physical appearances of the thing. *Function* refers to the intended use of the thing or the outcome the thing is supposed to produce.

Often transportation is viewed as not value-added because it doesn't seem to change fit, form, or function. However, when a customer wants transportation, it isn't a delay. In this case, transportation is a functionality characteristic. If the transportation does not get you where you want to go, then it is not value-added.

However, a company cannot meet customer needs if it doesn't exist. So, the relationship between the customer and the company must be mutually beneficial. Therefore, there are things that a company does for its benefit. These actions are called business value-added (BVA) and they meet three criteria:

- Business cares enough about it to pay for it.
- It physically changes the fit, form, or function.
- It is done right.

This means that there are three ways of being nonvalue-added[1] (NVA):

1. Neither customer nor business cares about it.
2. It doesn't physically change the fit, form, or function.
3. It has defects.

The definitions of value-added (VA) and nonvalue-added (NVA) apply to process *actions,* not to job positions or titles.

Either draw a map of the process actions (e.g., Figure 1.1) or document the actions by watching. Then, determine what type of action each is by using the criteria for customer value-added (CVA), business value-added (BVA), and nonvalue-added (NVA).

Remember not to label or call a position, job title, or resource (including people) VA or NVA. Only actions are CVA, BVA, or NVA for the purpose of determining root causes. In Figure 1.1, do not label Customer Service Rep as NVA. Instead, label action 2 as NVA because the information requested was also requested on the online reservation.

Delay Due to Slow Value-Added Actions

In theory, slow value-added actions exist only if all other delays have been eliminated and the process time is still unacceptable. One reason for tackling VA actions last is that the processes have not yet gone through delay reduction, and the NVA actions consume substantially more of a process time than the VA actions.

In practice, if after you have reduced NVA actions and reworked as much as possible (or as much as you want to) and the process time is still unacceptable, you can address the speed at which value-added actions are done. But, before you do, you should compare the amount of time for VA and NVA actions to be sure it's worthwhile.

One way to prioritize is to collect data on how long each CVA and BVA action takes. You can also ask those working the process to see if there is agreement on what are the top time-consuming actions.

Note that the customer and the business don't always want less time. Customers, for example, want concerts and vacations to last longer. Everyone wants to delay payments.

Besides replacing slower methods with faster ones, you can accelerate VA actions with standard operating procedures, continual training and practice, and using some advanced techniques, e.g., kanban (a method of inventory control).

How to Address Delay-Caused Defects

The second Lean principle says to create value through a value stream or process. To create more value, you should reduce as many nonvalue-added actions as possible.

The third principle is to make sure the value stream is always flowing. Stop only when a defect occurs—do not pass it on.

The fourth principle says that to create a smooth flow, pull rather than push. The objective is to never stop the flow of the value stream. But, you want the flow to be of value-added actions. Ideally, you would address delays in this order:

1. First work on solving delays due to nonvalue-added actions.
2. Then, solve delays due to flow stopping that is not due to defects.
3. Thirdly, solve delays due to defects.
4. Finally solve delays due to slow value-added actions.

One reason for this sequence is that removing nonvalue-added actions first simplifies the process so that it is possible that solutions for defects are identified and solved simultaneously. Even if you cannot identify and solve defects simultaneously, it will be much easier to find errors and other defect causes in the much simpler process. Another reason for removing NVA first is that there is no point in solving defects that result from these actions if you are going to remove them.

If the second step requires rearranging actions so that flow can occur, the rearrangement typically clarifies why actions are done. With fewer nonvalue-added actions and reduced rework, eliminating flow stoppage is also much easier to do. This revelation helps the third step be more efficient.

Now, with a much simpler process and with clarity on why the remaining actions need to be done, it is easier to address the defects. Nondelay defects can have four different causes. These are addressed in Chapters 2–5. You can use the same three-step procedure to address them:

1. Identify the type of problem by cause
2. Find the root cause
3. Address the root cause

How to Address Delay Due to Flow Stopping

Keep it flowing—With fewer nonvalue-added actions and reduced rework, eliminating flow stoppage is also much easier to do.

Creating value and keeping it flowing are supported by three techniques: Just-in-time (JIT) heijunka, and jidoka. JIT delivery creates flow, removes waits. Leveling demand and sequencing things appropriately reduces queues. This technique is called *heijunka*. The only exception to never stopping the flow is when there is a defect. The term *jidoka* refers to stopping the value stream on an abnormality rather than passing the defect to the next action.

Once the stops are identified, you can determine how to address the root cause by finding the answer to: What is it waiting for? Then reduce or eliminate what is being waited for. This could mean looking at this other process and the reasons for its delays. Note that the "thing" going through that process will be different.

Wait comes in three forms:

- Waiting while moving
- Waiting indefinitely (queuing)
- Waiting for definite times (batching, setting aside, or storing)

Table 1.1 also lists the ways to address the type of wait.

How to Solve Delay Due to Rework

Mistake proof—Rework is due to defects. If the defect is due to an error, you will learn how to address it in the next chapter. If the defect is not due to an error, read Chapters 3–5 to learn how to address these other defect types.

Note that legitimately reducing rework is the only direct way to simultaneously address in the right direction the three factors: quality (less rework means higher quality), cost (less rework cost), and time (less delay time).

Table 1.1 How to Address Waiting by Type of Wait

Type of Wait	*How to Address Waits*
Transportation, motion	Eliminate or replace with faster means
Storing	Just-in-Time (JIT) delivery, JIT processing
Queuing	Level demand, divide-and-conquer (e.g., checkout lanes for 7 items, 15 items, cash only, etc.)

How to Address Delay Due to Nonvalue-Added Actions

Eliminate—or, at least reduce the number of actions—Check that these actions are, in fact, NVA. Confirm customer and business requirements. Do a trial run. If confirmed that they are NVA, then eliminate them.

Another, often more effective way of eliminating NVA is to identify all CVA actions. Redesign the process with only these actions. Then identify all BVA actions; incorporate into the CVA-only process. The key to doing this is to never refer to title, positions, or people, only the actions themselves. For example, don't label Tom's or the clerk's position as CVA, but label each specific action that Tom or the clerk does as CVA, BVA, or NVA.

How to Address Delay Due to Slow Value-Added Actions

Replace with faster methods—After you have addressed as much as possible the three other types of delays, you can address the speed at which value-added actions are done. The solution is conceptually simple: Address slow work by replacing it with faster work. For example, email rather than fax, fax rather than regular mail. For slow repetitive work, use a machine rather than manual labor.

A second option to solving delays is, if irrespective of how many other delays you have, you discover a way of reducing value-added process time for an action and are willing and able to make the switch. Make the switch.

Delay Example 1

1) Identify Delay-Caused Defect

A Finance Department wanted to reduce the time it took for monthly closing. Reducing process time is a delay process problem.

2) Find Delay Root Causes

At a meeting to address the problem, the employees drew a map of the process flow of the things, people, information, and materials (also called a value stream map).

They also created a table with closing days as columns and the people doing the closing work as rows. They entered all the actions that needed to be done first. This revealed one problem: There was uneven distribution of work on the first day of closing.

This meant that there was Waiting. Everyone said they were busy. However, by looking at the templates and documents that needed to be completed, they agreed that they were waiting. In one case, the person doing the first essential work had more than a day's worth of work. The total workload was three people's worth, but only two people were working on closing the first day.

3) Address Delays

The department was willing to redistribute work to the five people in the department, resulting in more than half a day delay removed. This solution is heijunka—leveling work.

Delay Example 2

1) Identify Delay-Caused Defect

The good news for the company was double-digit growth in sales every year for the past five years. The bad news was that the capacity of the process between customer service and shipping had been exceeded. Or, so they thought. Ideally, after confirming an order with the customer, shipping needed the paperwork by 4:30 p.m. the day before the order could be shipped. More than one-in-four confirmations resulted in paperwork arriving after 4:30 p.m.

Shipping blamed Customer Service and Customer Service said that the volume was just too great to handle.

However, if they could reduce the time it took to process the paperwork, then capacity would naturally increase.

2) Find Delay Root Causes

The employees formed a team and collected some data (Table 1.2). They overlaid the data on their map of the process to help them find root causes (Figure 1.2).

Table 1.2 Occurrence and Duration Data for Types of Delay[a]

Type of Delay	*Delay Occurrence*	*Delay Duration*
Stopping flow (wait): Transportation	Customer Service to Scheduling and back Customer Service to Warehouse and back Customer Service to Shipping and back	Average of 108 minutes/ day in transportation (walks and sweeps) 17.1% of transportations are empty
Stopping flow (wait): Queuing	Scheduling Customer Service Shipping Warehouse	24 hours maximum 12 hours maximum 10 hours maximum 72 hours maximum

[a] Queues are such things as in- and outboxes or lines. Transportation moves product or documents. They include sweeps (a person travels to other locations to pick up items from outboxes at end of shifts) and walks (taking documents).

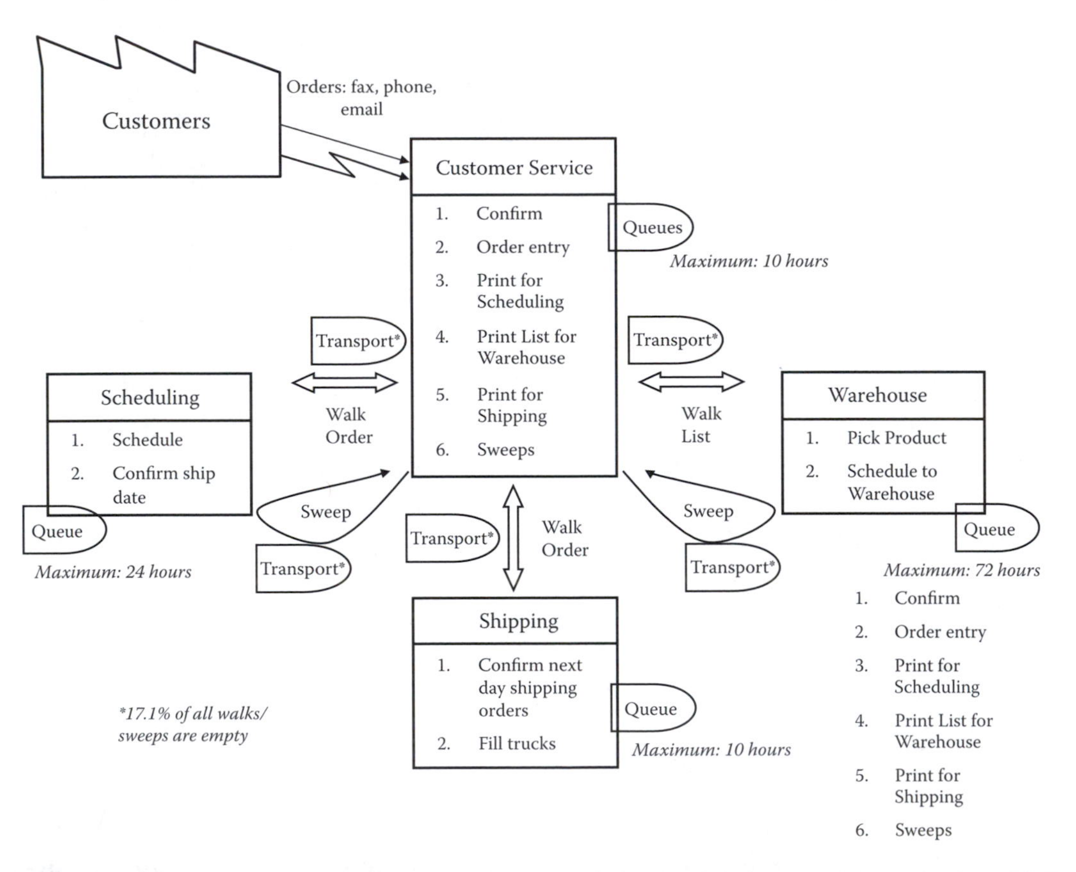

Figure 1.2 Delay Example 2.

3) Address Delays

Transportation—The company decided to plan for future growth by investing in an enhancement to the Order Entry software. They created screens that integrated Scheduling, Inventory, and Shipping information, thereby eliminating all walking to pick up documents. Two additional advantages were a reduction in paper and the ability to make customer-requested changes to orders faster.

Queuing—One of the pillars of the TPS is called heijunka, which means leveling and sequencing. The company leveled the demand by creating all paperwork at once starting a week earlier, which was easier with the software. This reduced duration of all four queues. They reduced queuing even more by having Shipping print the documents as needed rather than waiting for Customer Service.

The company took further action to prevent these problems from reoccurring. In other words, the company wanted to sustain its improvements.

Control to sustain performance—Managers monitored queuing at Scheduling, Customer Service, Shipping, and Warehouse using software that allowed them to track the actual times. Taking a cue from the airline industry, management established a "near miss" specification of 3:30 p.m., which monitored the percent of documents delivered between 3:30 and 4:30 p.m. Since their delay rate had been significantly reduced, they wanted a warning signal for potential problems. The 3:30 p.m. specification became the new definition of delay requiring preventive action when not met.

Delay Example 3

1) Identify Delay-Caused Defect

After meeting with prospective customers, field sales representatives create a proposal based on the customer requirements. That proposal gets sent to the regional office for approval. There are three regional managers who can approve the proposal. It varies whether one, two, or all three regional managers review and approve a single proposal. Sales reps have complained that they have lost some deals because, by the time they visit the prospective customer with approval, the customer has gone with a competitor.

In addition, to meet this year's financial goals, Sales has projected an increase in closed deals of 10% at current average value per deal. The conjecture for this project is that reducing approval time will help accomplish the financial goal and, thus, address the frustration of the sales reps.

2) Find Delay Root Causes

Data on root causes (Figure 1.3):

- Type of delays: Stopping flow waiting for transportation by overnight delivery.
- Where/when delays occur: (1) Between the submission of the proposal to the regional office and (2) the return of the proposal with a decision to the sales rep.
- Duration of the delays: Average 5.1 days.

3) Address Delays

Several sales reps and the office regional managers met to discuss two issues: (1) Whether there needs to be transportation and, if so, how can it be expedited, and (2) which handoffs are necessary.

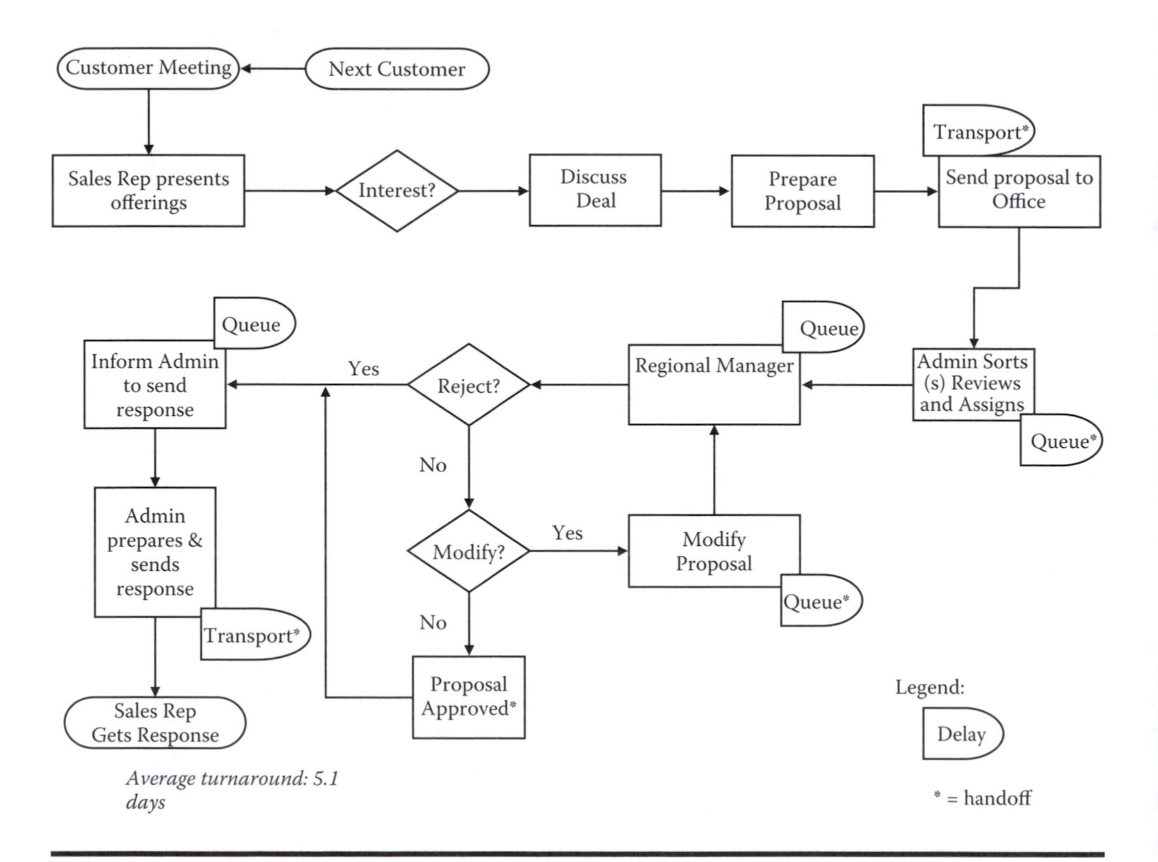

Figure 1.3 Delay Example 3.

The purpose of sending the proposal to the office was to get approval. The analysis showed:

- Five cases were rejected for reasons clearly agreed upon by all as deal breakers.
- Eleven approved cases met very similar conditions that the managers agreed were clear-cut approvals.
- In 31 cases, the regional managers felt they needed to review them more carefully before deciding because agreement was not reached quickly at the meeting.

Deal breakers would be completed on a one-page form with a red stripe across the top. The form was standardized and the condition that broke the deal was placed in a specific location on the form. Sales rep would send a copy to the regional office in an envelope with a red stripe.

Automatic approvals would be completed on a form with a green stripe across the top of each page. The form was standardized to include only those conditions that reflected the criteria for this case. The sales rep would send a copy to the regional office in an envelope with a green stripe.

All *other proposals* to the regional office were completed on a form with an orange stripe across the top of each page. These forms would go in irregularly sized envelopes with an orange stripe on the left side. The regional managers blocked every day from 10 a.m. to noon to review any orange stripe proposals that arrived that day. If none came that day, rather than cancel the meeting, they would review all proposals to ensure that they were being used correctly.

Control to Sustain Performance

The managers monitored the changes to see that they remained and were effective:

At their daily meetings, the regional managers reviewed nonautomatic proposals (orange-striped envelopes and forms). At these meetings, they kept track of exceptions to the color coding scheme, making sure that the forms and the criteria for deal breakers and auto approvals were adhered to.

They also kept track of the percentage of each type of proposal to see how well they were following their new procedures. They did this by determining whether any orange proposals took longer than one day.

What You Can Do Now

Every process takes time. It's rare that processing time cannot be improved by removing one or more of the four types of delays. By identifying and addressing these delay types, you should be able to reduce your processes' times.

The fifth principle is to seek perfection by continuously improving even if it is only incrementally. Over time, the incremental improvements accumulate to a significant gain.

For cases that are more complicated, there are advanced tools and concepts that are useful. Learn about them (JIT learning) when you encounter a process whose cycle time you cannot improve with what you know.

There are five things you can do to acquire the habit of identifying delays and gain the expertise in reducing and removing delays:

- Go to the gemba. Everywhere you go—the store, airport, online, at home—watch processes and practice identifying the four types of delays.
- Practice reducing delays on processes you do at work and at home (e.g., loading the dishwasher so that unloading takes the least amount of time).
- Partner with others at home and at work to help each other reduce process time.
- Study tools used to reduce motion, create continuous flow, and create pull, e.g., 5S, cell design, quick changeover, kanban, single piece flow, the supermarket.
- Teach others, review the training chapters (Chapters 11 and 12) and create simulations so that others can learn by doing.

Okay, go reduce delays.

Endnotes

1. To help identify what is NVA (nonvalue-added), TPS uses seven wastes that can be remembered by the acronym TIMWOOD:
   - Transportation
   - Inventory
   - (Unnecessary) Motion
   - Waiting

- Overprocessing
- Overproduction
- Defect

Transportation, Inventory, Motion and Waiting (TIMW) are nonvalue-added because they do not physically change fit, form, or function. Not physically changing fit, form, or function is what is meant by stopping the flow. However, there is value-added transportation when it is what the customer is paying for, e.g., flight, taxi, bus.

Overprocessing and overproduction (OO) are activities that physically change fit, form, or function. But, they are labeled “over” because the customer does not want them and is not willing to pay for them. Overprocessing is adding unnecessary features or enhancing a product or service unnecessarily. Overproduction is creating more than the demand, which typically results in capital set aside with diminishing value over time. Because it takes time to do either of these, it results in delay.

Defect (D) is an example of NVA because the customer neither wants defects nor wants to pay for the defect or the rework associated with it.

Overprocessing, overproduction, and producing defects are examples of nonvalue-added action being taken on the thing.

The seven wastes do not include delays due to the fourth reason: slow value-added activity.

Chapter 2

Error-Caused Defect

How to Identify Error-Caused Defects

"This isn't what I ordered..."
"This isn't the right information…"
"They forgot…"
"It's not the right color (size, amount, …)"

We all make mistakes. We all recognize when a defect occurs caused by a mistake or by an error.

For example, incorrect information on documents, such as an invoice with incorrect address, name, amount, or quantity, is due to error. It need not be a person who made the error. A machine might have a software miscoding that searches for the product code to determine the invoice amount.

Looking at it in reverse also can help determine whether an error is the root cause. Does the process include sorting, identifying, labeling, writing, entering, interpreting, coloring, selecting, communicating, moving, etc.? If the result (sort, identification, label, writing, etc.) is wrong it could be caused by an error. It doesn't matter whether it is done by a person or a machine.

If you are calling a defect an error (or mistake or wrong) rather than a defect, then you probably have an error type of process problem. Once you know the error, you know the root cause. You just don't know where it occurred.

Key Words: accuracy/accurate, mistake, error, wrong.

How to Find Root Causes of Error-Caused Defects

Retrace the process—Start from where the error is noticed and then retrace to the origination of the information. While retracing, note all the touch points, actions in the process that manipulate the item or information.

For example, you order six items, but get only five. There is a defect of the wrong amount. You might say upon counting them: "There is an error (or they made a mistake or this is wrong), I ordered six, but there are only five."

Identify all the actions of receiving, documenting, passing, or using/manipulating the information about how many you ordered. Only those actions that manipulate or use the information can have an error.

- Was the error that someone packed five rather than six?
- Was the error documenting five on the order form rather than six?
- Was the error that you mistakenly said five rather than six when you ordered?
- Was the error due to some other action at another touch point?

If there was more than one such action, you may not necessarily determine which caused this specific defect of wrong amount. You could only determine that if you knew (e.g., by documentation) before and after each action what number was ordered.

For example, assume you called to place your order. This is an action that involves verbally passing the order quantity and, therefore, is subject to error. Unless you have documentation that you gave the right number and the person receiving your order recorded or heard the wrong quantity, you won't know the error occurred then.

Sometimes you will be able to find the specific action that caused the specific defect. You will have an order form that shows that you ordered six items. However, that information had to be entered into an order system and you see that the amount in the system is five. Then you know the error was order entry.

Note that some actions might appear to be a single action, but when detailed you realize it is more. For example, the touch point where you place the order might be viewed at a high level as "place order." However, if you and I miscommunicate the quantity, is the error that you mistakenly said five or that I misheard you say five?

Be aware that transfer of information involves more than one action and you need to identify all the actions where errors can occur.

We may not agree on which specific action was the cause, but we do know that it was miscommunication. We know when and where it occurred.

How to Address Error-Caused Defects

Mistake-proof the process—We all know examples of mistake-proofing. We try to enter a card in a slot, but it only goes in one way. We enter a credit card number on an online form and it rejects it because we mistakenly entered the wrong digit or insufficient digits.

Errors are best addressed by mistake-proofing. The concept of mistake-proofing and its tools comes from the Toyota Production System (TPS), where it is known as poka yoke. Mistake-proofing and other interventions are described using the following sequence:

$$\text{Error} \Rightarrow \text{Defect} \Rightarrow \text{Consequence}$$

You can intervene in the order from worst to best:

a. After the consequence
b. After the defect, but before the consequence
c. After the error, but before the defect
d. Before the error

Case (a) requires mitigating the results, e.g., the insurance paying for damages. Cases (b) and (c) are typically combined with 100% inspection before (a) and (b), respectively. However, inspection will not catch all errors and defects. Case (d) is the ideal intervention: mistake-proofing.

Audit your intervention—even audit mistake-proofing.

It is not always possible to mistake-proof, but to whatever extent you can intervene before the consequence, you reduce the probability of errors, resulting in fewer defects and reduce the probability of someone suffering the consequences.

Even when you do not know which specific action caused the specific error, you can still improve the process. If all you do is try to mistake-proof the action that resulted in this defect, the process will still be prone to errors at the other actions.

For continuous improvement, continue mistake-proofing as many of the actions that could have been the root cause until they are addressed. In fact, you do not even have to wait for errors. You can identify possible defects due to errors right now in your processes; retrace each specific defect to the specific actions that could cause that error and mistake-proof them.

If you can't address all the actions, then prioritize them. Consider such factors as the impact of the consequences, the cost of intervention, the time it takes to intervene, and so forth. For example, is the wrong amount worse than a misspelled name or a wrong address?

Error-Caused Defect Example 1

1) Identify Error Defect

In Figure 2.1, you are the analyst that does the action: Create Monthly Invoice with Rate from Document. You receive a loan document from the third-party lender and notice that the rate is wrong (defect). You realize this is a defect due to an error.

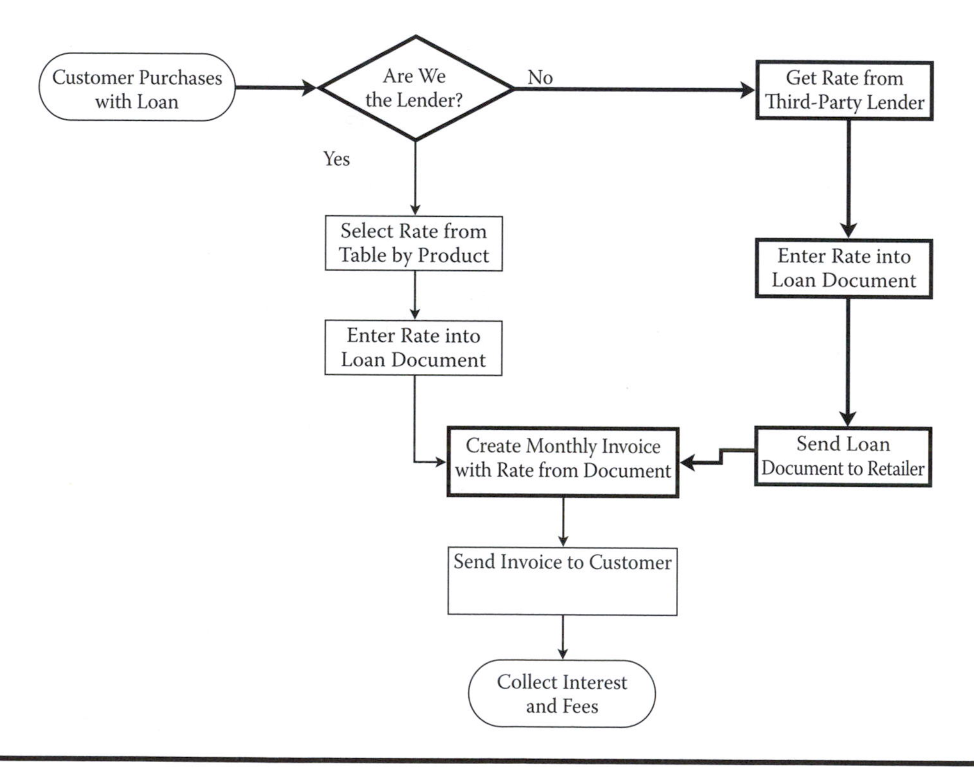

Figure 2.1 Error Caused Defect—Example 1.

2) Find Error Root Cause

You retrace the process. Looking at the previous actions (bold boxes and diamonds connected by bold arrows), there are only two touch points: (1) getting the rate from the third-party lender and (2) entering rate on the loan document.

At the first touch point (1), it could be that the person from the third-party lender communicated incorrectly or the person receiving the rate heard it incorrectly. Unless the communication is recorded or either person recognizes and admits the error, you cannot know which it is.

At (2), only writing the wrong rate on the loan document can be the error.

Rather than spend time trying to determine which error caused the defect this time, you decide to address all the possible errors at the touch points.

3) Address Error

Rather than using people to communicate and record, you propose having (1) a database that finds the rate based on mistake-proofed information entries and (2) emails or faxes the loan document with the rate to you, the analyst.

This pushes the possibility of errors to the creation of the database and subsequent changes. However, if that information is correct, then the three possible errors are eliminated. In addition, two communications also are eliminated and only one person is involved rather than three. The cost of this technological mistake-proofing solution would be the determining factor.

Until this solution is implemented, you implement less effective solutions to include, but are not limited to, the following:

- Record rate communication at first touch point.
- Have rate provider email confirmation of rate after initial communication.
- Replace handwriting on loan document with double electronic entry.

Error-Caused Defect Example 2

1) Identify Error Defect

In Figure 2.2, the customer notices that the total dollar amount is wrong on the invoice, which means an error occurred somewhere.

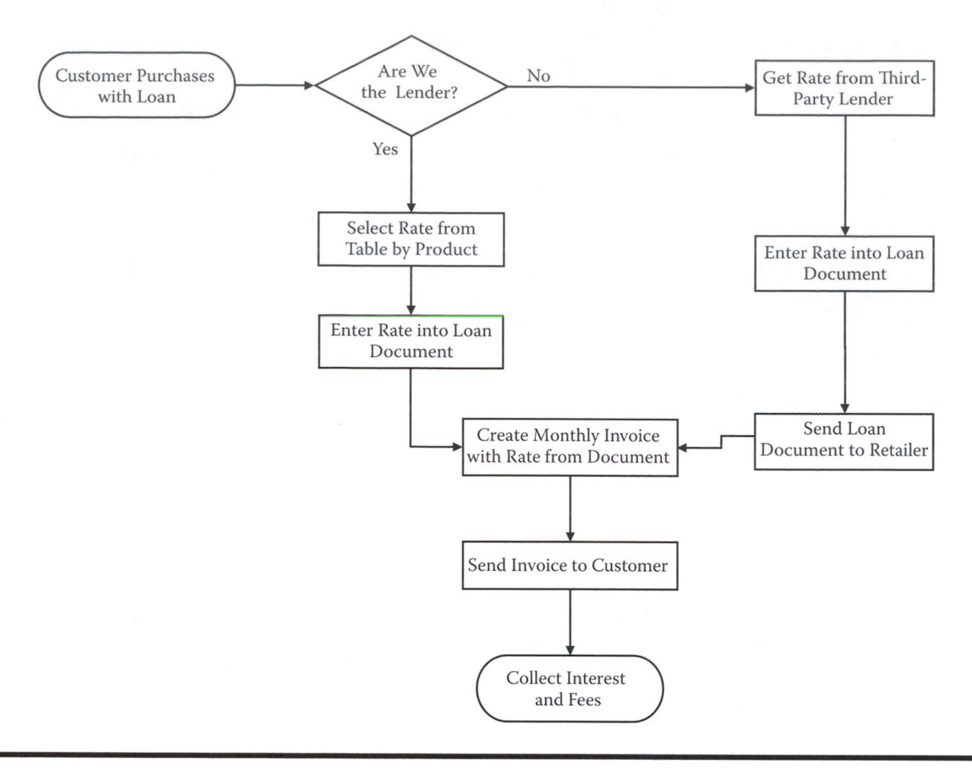

Figure 2.2 Error Caused Defect—Example 2.

The problem is more complicated than Example 1. Is it the wrong rate or other areas that make up the total dollar amount? If it is the wrong rate, did it occur with the third-party lender or within your own company?

2) Find Error Root Cause

Retracing, you see that not only can the error occur at the two touch points identified in Example 1, but it also can occur at three other actions: (3) selecting the wrong rate from the table, (4) entering the wrong rate into loan document, and (5) creating an invoice with the wrong total amount.

3) Address Error

Sometimes you can fix several actions by replacing them with only one improvement.

Mistake-proofing these three errors could involve a technological solution where a loan document identifier is entered into an electronic form with double-entry. The software then selects the rate, enters it onto an electronic loan document, calculates the total invoice amount, enters it into the invoice, and prints the invoice. It could even send the invoice to the customer.

Less effective interventions would be similar to before, e.g., double-electronic entry.

What You Can Do Now

Without mistake-proofing, every process that uses people at the touch points will have errors. Now that you have read these few pages, you can improve almost any process because defects due to errors do occur that frequently.

There are several things you can do to acquire the habit of finding error root causes and of mistake-proofing:

- Practice identifying defects caused by errors at home and at work; when you create a new process, identify where errors may occur for each piece of information.
- Practice retracing to the touch points even when the process has no errors; ask yourself: "Could this be wrong?" and "Where could the error occur?"
- Study the mistake-proofs that exist in your everyday life, both for human and mechanical error; determine the error the mistake-proof prevents.
- Practice mistake-proofing things you do at work and at home (e.g., how would you arrange dishes in a dishwasher, clothes in a drawer or closet, tools in a drawer or garage so you can correctly select each using a blindfold?).
- Learn more about mistake-proofing; do a Web search for mistake-proofing and poka yoke.
- Understand more by teaching others and working on each other's processes.

That's all you need to know. Go reduce errors.

Chapter 3

Suboptimality-Caused Defect

How to Identify Suboptimality-Caused Defects

"One of the six I ordered isn't the same shape."
"These two parts just don't fit quite right."
"Every once in a while, it just doesn't work."

Problems of fit, form, or function are typically suboptimality problems.

You order two special batteries for your digital camera. If you only get one, you know it is an error problem. If it arrived a week later when you selected two to three business day shipping, you know it is a delay problem. But, how do you suspect it is a suboptimality problem? It is if:

- You try the battery and it doesn't work.
- You try to insert the battery and it doesn't fit or you can't close the cover.
- It has a defect like a scratch or blurry writing, e.g., the brand name.
- You are calling it a defect rather than an error.

The assumption that all processes have variation is the basis for this third type of process problem. Two circumstances can occur. One is that the amount of variation in the process is so large that it will always produce some defects even when no errors are committed. The other is that the settings for running machines or standards for raw materials are off enough that it will always produce some defects even when no errors are committed.

When one or both of these two circumstances occur, the process is not running optimally.

A manufacturing process typically takes raw materials and converts and assembles them into a finished product. If the raw materials do not have certain characteristics or the machinery is not functioning optimally, then the resulting product has defects, including cosmetic flaws, ill-fitting parts, and lack of functionality. It may not be obvious which settings are critical and what their optimal levels should be. This is the case where Six Sigma's statistical approach is quite beneficial.

If one is willing and able to control these factors (machines at their optimal settings and raw materials with their optimal characteristics), then one can use statistical analyses (hypothesis testing and designed experiments) to reduce the frequency of defects.

Suboptimality does not just apply to manufacturing where machinery settings and characteristics of raw material are critical, it also applies to procedures and, therefore, to services. In the context of procedures, the lack of optimality might be called substandard procedures. There might not be any machinery or raw material involved, yet a substandard or suboptimal service is provided.

The statistics useful in efficiently finding these optimal characteristics for raw materials and optimal settings for machinery is also applicable to nonmanufacturing processes or products. They have been used for designing forms, determining marketing campaigns, creating recipes, increasing crop yield, and many other service processes.

For example, you are trying to fill an empty position. Your hiring process is completed in the time you had planned, but when you offer your first-choice candidate the job, he declines it. There is no error and there is no delay. This could be a suboptimality problem.

Key Words: scrap/waste, quality, rework, capacity, uptime low/downtime high, chronic or sporadic or intermittent failure, can't make product, unplanned maintenance, broken (process, step, machine, component, part), doesn't work.

How to Find Root Causes of Suboptimality-Caused Defects

For errors and delays, you did not actually have to collect data to find root causes. Rather than looking everywhere for where an error or delay occurred, you could significantly reduce where you search. You may not have known which actions caused a specific defect, but, you knew that errors causing the defect could occur only at those actions. You knew that delays could occur only because of the four reasons cited on page 4 (stopping flow, repeating flow, nonvalue-added actions, slow value-added actions).

Without collecting data, you could make changes to the process knowing improvements would result. For example, if you reduce the time to transport or wait, the total time (everything else being equal) will be reduced. If you mistake-proof a credit number entry, then the probability of charging a nonexistent or expired credit card is zero.

However, when it comes to suboptimality problems, there is no getting around data collection and data analysis.

Because you do not want to collect data on every possible factor, you need to use methods for selecting plausible factors. Such methods include scientific laws and theories, historical data, and data-based experiences.

The more you rely on subjective experiences and opinions, the more likely you will increase the number of factors studied and include factors that are not root causes.

Once you have determined the factors on which you need to collect data, then you must design the study or experiment to collect these data. There are numerous books on design of experiment (DOE) ranging from simple to complex. The next sections covering DOEs, data collection, and data analyses highlight some key issues in determining causal factors.

Design of Experiment (DOE)

The simplest experiment has one factor. Statistical experimental designs require at least two factors because the one-factor case is straightforward. However, the one-factor case can help you understand some practical aspects of data collection and analysis.

Suppose you are taking a class and someone complains about the room temperature. There is a thermostat that controls it, but what is the temperature you set?

Typically, one person adjusts it. Eventually someone else complains that it's now too hot or it's too cold. The thermostat is adjusted again. This happens several times.

This approach is called *trial and error*. No one measures the effect for each adjustment, so no one can determine the optimal setting.

A DOE would tell you what settings to choose to collect data and then the analysis of the data determines the optimal setting. This approach assumes you have a measure that you want to optimize.

For example, if you want to maximize the number of people who are comfortable, then you might select three settings and at each setting count the number of people who are comfortable. The analysis determines what setting maximizes that number, even if it wasn't one of the settings studied.

There are two other considerations when trying to determine causation. One issue is confounding and the other is interactions. These issues can be managed by controlling them and randomizing how you collect data.

Let's review the thermostat example to understand these two issues. You have decided to test three temperature settings: 67°F, 70°F, and 73°F. Suppose the room has big windows and the morning is cloudy while the afternoon is sunny. If you set the thermostat at 73°F in the morning and 67°F in the afternoon, you might not feel a difference.

This is an example of confounding: two factors change at the same time so you don't know which is the cause of the result changing. You can reduce the probability of this occurring by randomly testing the different thermostat settings.

You might choose to test each thermostat setting twice during the day. You would randomly determine the order in which you set the thermostat. That reduces the probability of a factor you do not control from being confounded with the factor you want to study.

An interaction occurs when you need two or more factors to get the result you want. You can view the cloudy–sunny conditions as a factor that interacts with the thermostat setting. One way to neutralize this factor is to use shades or blinds.

However, if you cannot do that, then you would want to know how much of an effect sun has on a thermostat setting. For example, if 71°F is the desirable setting when there is no sun, then an interaction would mean setting the thermostat at, say, 68°F when it is sunny.

If you do experiments without controlling nuisance factors and randomizing your data collection plan, then you might have confounding variables and you won't understand the interactions.

Data Collection

You want to determine the specific causes. The key concept behind the way data are collected for suboptimality problems is to control the experiment. Correlation does not mean causation. To ensure causation, you must control the experiment.

If you use historical data where the process was not controlled for the factors you want to study, the best you can hope for is correlation. In addition, you will have two other issues when analyzing the data. First, you will not be able to recognize if there are interactions. Interactions occur when the optimal level for one factor depends on the level of another factor. And, secondly, if two factors change simultaneously, then you will not know which is the root cause. When this occurs, the factors are said to be confounded.

With DOEs, you avoid all three issues: (1) not showing causation, (2) not detecting interactions, and (3) not having confounding. You will have controlled the experiment, you will understand the interactions, and you will avoid confounding.

The simplest DOE is to look at all possible combinations of the factors you are studying. This is called a full-factorial DOE. It tells you everything about these factors without any of the three issues.

To control the experiment, you have to select the factors and, for each factor, select the settings or levels you will study. For example, you might select temperature as one factor with 100°F, 150°F, and 200°F as the levels, and another factor as time, with 5, 10, and 15 minutes as the levels. Then, you decide for which combinations of these factors and levels on which you will collect data. There are 3 x 3 = 9 different combinations of these two factors with three levels each. For example, you might collect data at temperature 200°F for 10 minutes.

All other factors are ignored or held constant. To ensure that these other factors do not "intrude" on the study, you randomly run the process at the combinations of the study factors and levels. This requires more effort because you have to change the process each time you collect data for other combinations. This is one reason why DOEs are often not run. However, the gain can be substantially more over time than the cost and time of doing such experiments. Also, there are ways to reduce the amount of data collected.

By making some assumptions, you can do a study that doesn't look at all possible combinations. These DOEs are called *fractional factorials*. Fractional factorials, however, do have confounding variables, but you already know this and can study it. These DOEs are more complicated to analyze and beyond the level of this book.

Often you conduct more than one study. One reason is that there is no guarantee that the factors you select to study include the root cause(s). You will have to generate more factors to study, and then collect data and analyze them when this occurs.

Another reason is that when you do a fractional factorial and the root cause is in the confounding variables, you will need to do another study to resolve these. However, this will be a smaller study and one can plan so that the total amount of data one collects are less than if one were to do a full factorial from the beginning.

Data Analysis

DOEs can be analyzed logically. Typically, however, the analysis is done statistically using software that shows the results graphically and numerically with some statistical assumptions.

The results state which factors are statistically significant as root causes. A factor can be significant individually or in combination with others—this is an interaction. In addition, the analysis will determine what settings the significant factors should be at for optimal performance. This performance is optimal relative to the levels studied.

When you are done with the analysis, you will know both the critical (or root cause) factor(s) and its (their) level for optimal process performance.

Unfortunately, it is possible that the optimal performance of your process is not good enough. For example, the optimal performance of regular mail will never be as good as email. In such cases, you know that you have to replace the process or overhaul it.

How to Address Suboptimality-Caused Defects

Set and control/maintain your critical factors to their optimal levels.

Typically, solving the problem is substantially easier than identifying the root causes. Identifying root causes means identifying the factors that are critical and the levels for those factors. The levels become the specifications for the factors. Solving the problem means controlling the process to keep the factors within the specifications.

Typically, the specifications will have a range, e.g., 10 minutes ± 1 minute. When this is a range, the lower value is called the *lower specification limit* (LSL), the higher value is called the *upper specification limit* (USL), and the ideal value is called the *target* or *nominal*.

If you had studied temperature and time and found that 150°F for 15 minutes was optimal, then you just need to set the temperature accordingly (within its range) and make sure you run for only 15 minutes

(± the allowable deviation). Sometimes this is easy because there are dials and levers and software that can be "told" this information.

Other times, it is not as easy. For example, the environment may affect temperature so that the summer setting at 150°F will result in a higher actual temperature, perhaps exceeding the allowable deviation.

Suboptimality Example

1) Identify Suboptimality-Caused Defect

A manufacturer of metal fasteners requires that the opening of the fastener be a specific amount 0.08 in. to ensure the fastener functions: not too loose and not too tight. The LSL = 0.045 in., the USL = 0.112 in., and nominal = 0.081 in.

They currently were having problems with too many fasteners having too narrow an opening and being scrapped. If the opening was too wide, they sometimes could rework it, but the cost of manual labor to reprocess was getting prohibitive.

This was not an issue of delays or errors. They assumed the setting of the machinery was not optimal as some parts did not fit even when the machine was set according to procedure.

2) Find Suboptimality Root Causes

To search for plausible root causes, the operators collected historical data on two factors (Table 3.1) that they were willing and able to control.

Recall that historical data can only show correlation. Correlation simply means that as one factor changes so does another. The results (Figure 3.1) show that surface hardness (Figure 3.1a), air pressure (Figure 3.1b), and/or

Table 3.1 Description of Factors in a Design of Experiment (DOE)

	X1	*X2*
Name	Air pressure	Surface hardness
Definition	Pressure (psi) to advance fastener on moving bed	Bedding surface
Levels	Three levels (groups): 37–41, 42–46, 47–51	Two levels: Soft, hard

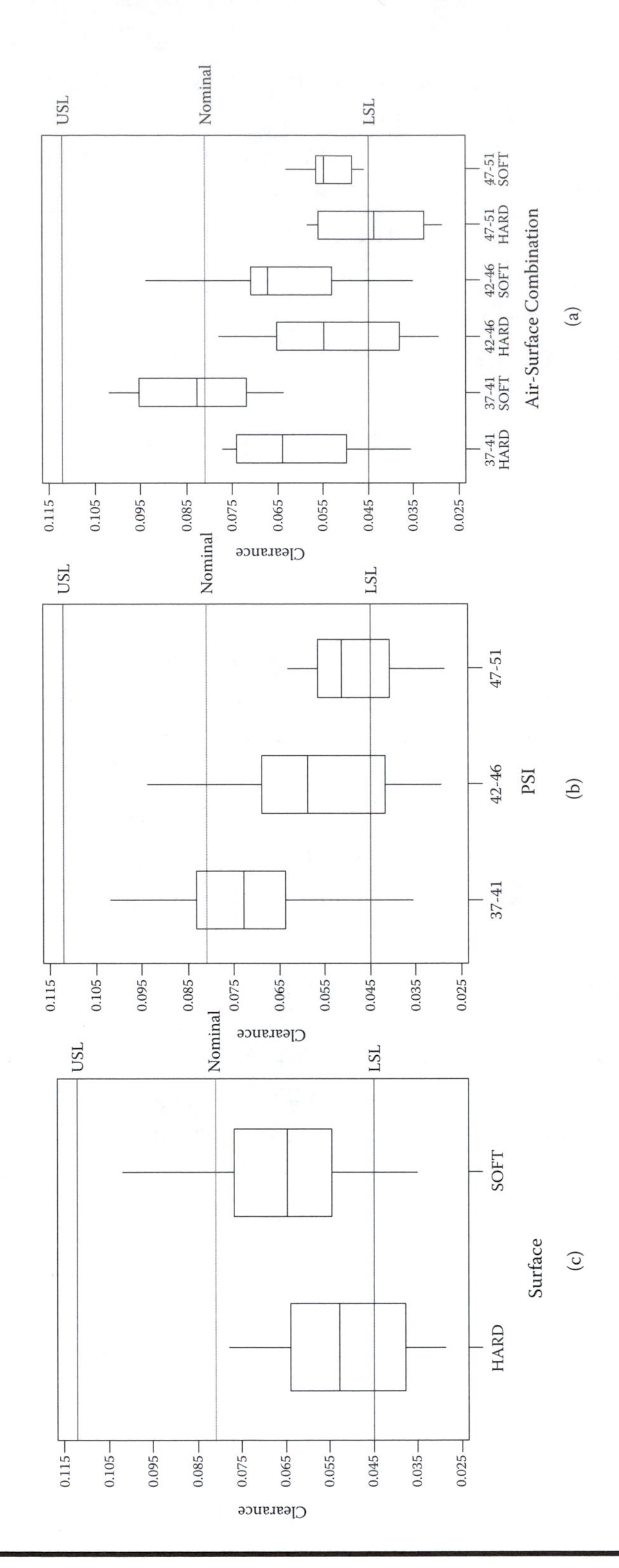

Figure 3.1 Historical Analysis for Suboptimal Setting.

their interaction (Figure 3.1c) were significantly correlated with three types of performance:

- Average clearance (they could move the distribution mean closer to the Nominal of 0.081 in.): in Figures 3.1 a–c as the horizontal axis factor changes there are changes on the vertical axis comparing the lines in the middle of the boxes.
- Variation (they could reduce the range of clearance results, making it less likely to be outside of specifications): in Figures 3.1b and 3.1c as the horizontal factor changes, there are changes on the vertical axis scale as measured by the size of the boxes.
- Percent good (meeting specifications) (they could combine the better average with less variation): Figure 3.1a as the horizontal axis factor changes, there are changes on the vertical axis as measured by how much of the box is between the LSL and the USL.

While the combination 37–41 psi and soft was significantly less variable than any other combination, its average was too far from target to help increase percent good. The best combination for percent good appears to be 37–41 psi air pressure and a soft bed (Figure 3.1c). Also, they knew that the historical data only showed correlation.

Because there is only one factor, they did a DOE on air pressure, holding the soft bed constant. They tested five pressures: 37, 38, 39, 40, and 41. Figure 3.2 shows that variation and percent good (all values are within the specifications) were not an issue in this psi range.

Table 3.2 shows that there was no difference in means in the range 38–41 (the letter B), but that 40 and 41 were significantly lower than the other psi (letter C) while at 37 psi, it was significantly higher than at any other psi (letter A). Because the center of the specifications is nominal = 0.081, the company set the psi at 39 ±1. They predicted that they could get 99.92% good.[1] Levels not connected by the same letter are significantly different.[2]

3) Address Suboptimality

The easy solution is to set the bed surface lining to soft and the air pressure to 39 psi. A more rigorous solution involves more considerations to ensure that this solution will be permanent and is robust to other uncontrolled factors.

The company replaced all hard surface linings with soft linings, specifying the fabrication needed. They purchased an instrument to measure

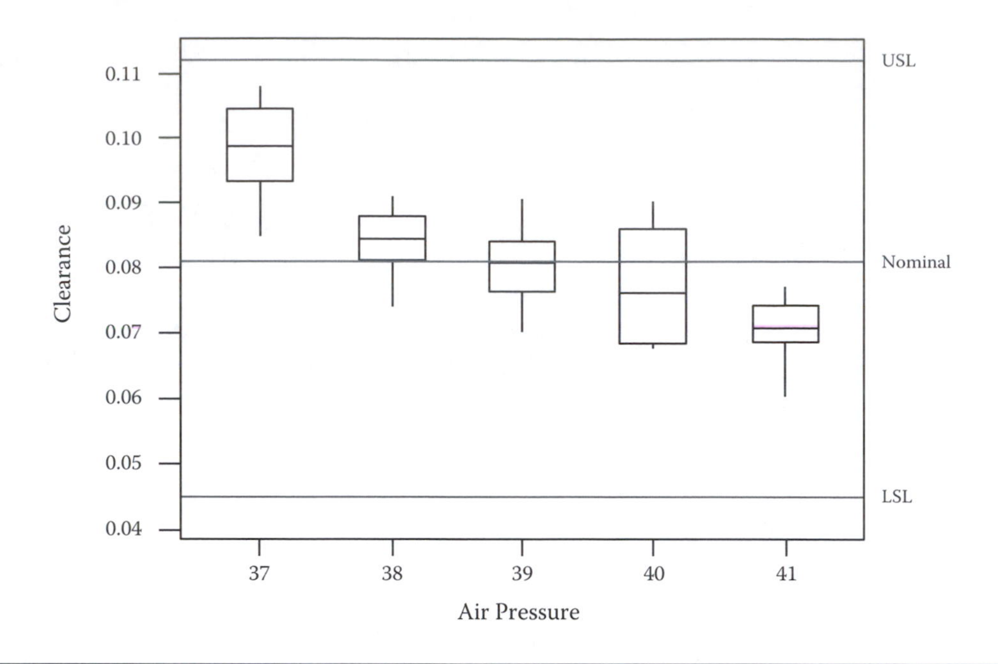

Figure 3.2 DOE Analysis for Suboptimal Setting.

Table 3.2 Results of DOE Indicating Significant Factors for Optimizing Air Pressure

Level	*Statistical*	*Significance*		*Mean*	*Standard Deviation*
37	A			0.0984	0.007096
38		B		0.0843	0.004943
39		B		0.0812	0.006351
40		B	C	0.0777	0.008354
41		B	C	0.0704	0.004835

the softness, allowing more accurate differentiation between the soft and hard and created specifications for softness. They set a schedule of testing the softness once a day, and then extended the time between tests until they had enough data to set up regular maintenance of the bed.

The air pressure specifications were set at 39 ± 1 psi. They located a gauge at the site where the air pressure would be verified and documented. They instituted a procedure to verify the air pressure during and after shifts and whenever there were stops. They then added another procedure on when and how to adjust the air pressure when it was not within specifications.

What You Can Do Now

Every process has variability. If the reason for defects is one of variation (it doesn't matter whether any errors are made) and you still get some defects, then you probably have a suboptimality problem.

These problems are typically solved by or with the help of statistical expertise because they do not occur frequently enough for someone to quickly acquire the skill. The three-step problem-solving procedure involves more work: designing the experiment, running the experiment, collecting the data, analyzing the data. However, its use can help you avoid having to regularly solve these defects.

If you are in a function, e.g., R&D, QA, manufacturing, engineering, testing, then your job description may require or at least benefit from skills in solving suboptimality problems.

There are four things you can do to acquire the skills to do formal studies and have the expertise in finding critical factors and optimal settings:

1. Learn more about experimental designs from both a logical and statistical view, e.g., design of experiment (DOE), Kepner–Tregoe, Shainin; search the Web for case studies on DOE.
2. Study tools used to plan, do, and analyze experiments including software, e.g., MiniTab® and JMP®.
3. Practice using these tools on processes you do at work and at home (e.g., a DOE on a recipe for an entrée, dessert, or grilling sauce), and on products or processes you have already created, similar to a postmortem.
4. Learn how to set specifications for the root cause factors and control them (e.g., determining references so that you always pull your car into the garage optimally to the same spot).

To teach others, review the training chapters (Chapters 11 and 12) and create simulations so that others can learn by doing.

Endnotes

1. Combining the clearance values for psi 38, 39, and 40 yields a mean = 0.0876 and standard deviation = 0.00716. Using normal distribution theory, this gave a lower percent good of 99.92% with 95% confidence.
2. This analysis is a comparison for all pairs using Tukey–Kramer HSD (Honestly Significant Difference). This is an advanced analysis that occurs in many DOE software.

Chapter 4

Unpredictability-Caused Defect

How to Identify Unpredictability-Caused Defects

"We weren't expecting this demand and now we're backlogged."
"Management is suggesting rebates or discounts to reduce our inventory levels."
"What's the point of forecasting if it's never accurate?"

The sequence below describes how this process problem occurs. You identify predictors that you think will tell you what the outcome will be. You check the levels of these predictors. You insert those levels in a model. You make a prediction. Then you compare your prediction to what actually happens. If the prediction and reality do not match sufficiently closely, then the root cause is the model.

Unpredictability sequence:

Predictors → check actual predictor levels → model → prediction → comparison to reality

The key difference between this case and suboptimality is the second step. In suboptimality problems, the factors are the root causes that you can *control* to cause the outcome. In the prediction model, the predictors cannot be changed or controlled. You can only predict the outcome.

The root cause of an unpredictability defect is a model with uncontrollable factors. The defect in the prediction is not ever accurate to the level desired or accurate frequently enough to the level desired. You can change the model or change the predictors to try to improve the predictability, but you cannot change the levels of the factors.

It is possible that there isn't a model that does any better or that even predicts at all.

For example, there are medical treatments that do not cure 100%, but at a lesser percentage. Still, you use these treatments because the success rate is better than not using the treatment at all.

This is similar to the suboptimal settings case, but you stop at correlation. You can't control the predictors. In most cases, you can't control them because of physical barriers or you ought not to control them because it's illegal. For example, a company can change prices to influence customers to buy their products and services. However, it should not conspire with competitors to fix prices or use force to get customers to buy its products.

The use of statistical tools and software are essential for unpredictability problems. It might appear that this type of problem is rare. However, every company depends on making predictions: predictions in finance (forecasting), production (demand and inventory), Human Resources (HR) (employee contribution, longevity, resources), marketing (viable product or service), Information Technology (IT) (disk space, downtime, service calls), risk (financial and product, e.g., stability, durability), and so on.

Because the purpose of prediction models is to have accurate predictions, two requirements are accuracy and timeliness of the prediction. For accuracy, you will need to state the amount of deviation that is acceptable as specifications.

Key Words: customer demand (variable, high, low), too many/few (products/types, people, resources including people), schedule (variation), inventory (off, low, excessive).

How to Find Root Causes of Unpredictability-Caused Defects

Correlation Analysis—Because you are not looking for controllable root causes but predictors for unpredictability problems, the search can include nonplausible as well as plausible predictive factors.

Plausible factors are those factors that are indirectly connected to the result to be predicted. They, in certain conditions, can cause the result, cause something else that causes the result, or be caused at the same time as the result. Even if they are the cause, they differ from the suboptimality causes because you cannot or should not control them.

Nonplausible factors are those that serendipitously correlate with the result. Statistical correlation analysis will determine whether there is a correlation and whether the predictions are accurate enough to be useful. You can find statistical analyses and software for a variety of prediction models on the Web.

How to Address Unpredictability-Caused Defects

Create a procedure on how to use the predictors—Typically that means collecting information to determine what the current values are of the predictors used in the model. Then, either manually or using software, plug in those values and calculate the outcome. That is your prediction.

Because the model is a correlation model rather than a causal model, you should continually verify the model's usefulness, especially with serendipitously correlated predictors. Therefore, the procedure should also include (1) criteria for when to reassess the model and predictors and (2) how to reassess the model and predictors (e.g., when accuracy drops below a certain level).

Unpredictability Example

1) Identify Unpredictability-Caused Defect

A bank can increase its revenue by loaning a greater amount to a customer. One customer segment (small businesses) has a credit line from which it borrows money. If the customer uses more of the credit line, the bank makes more money.

A typical use of the credit line was for payroll. The bank decided to take the daily average of credit line utilization per calendar week. The bank wanted at least 85% utilization. This was the lower specification limit (LSL).

Because the bank cannot cause the customer to use the credit line, the bank started a project to determine what factors might be predictors of more credit use. It might be able to influence those factors.

2) Find Unpredictability Root Causes

Analysts identified the following factors shown in Table 4.1.

Table 4.1 Description of Factors for Analysis

Label	*Prediction Factor*	*Definition*	*Choices*
X1	Credit Line	What is the limit of the credit line (in $000000)?	0.5–1, 1–2.5, >2.5
X2	Referral	Was the customer referred to the company?	Yes, No
X3	Assigned Contact	Does the person have an assigned company contact?	Yes, No
X4	Accessibility	Does the contact forward calls to his/her cell when not at the desk?	Yes, No
X5	Informed Customer	Does the company inform the customer of credit availability?	Yes, No
X6	Referral Utilization	Average credit utilization of referral	1–25%, 26–50%, 51–75%, 76–100%
X7	Contact Frequency	How often is customer contacted?	0, 1/week, 1/month

The team developed a model fitting a percent credit utilization to the seven predictors. Figure 4.1 shows which value of each significant factor maximizes credit utilization.

Prediction Model

Table 4.2 shows the results of the statistical test where a superscript in the "P" column indicates a statistically significant correlation. Removing Credit Line, Informed Customer, and Contact Frequency resulted in Referral Utilization not being significant. Combining the analyses, the team concluded that Referral, Assigned Contact, and Accessibility were the critical predictors.

Table 4.3 shows the results of the final analysis with the prediction equation:

Predicted Percent Credit Utilization = 83.18 +/– 4.04* Referral +/– 3.04* Customer Contact +/– 1.90* Accessibility

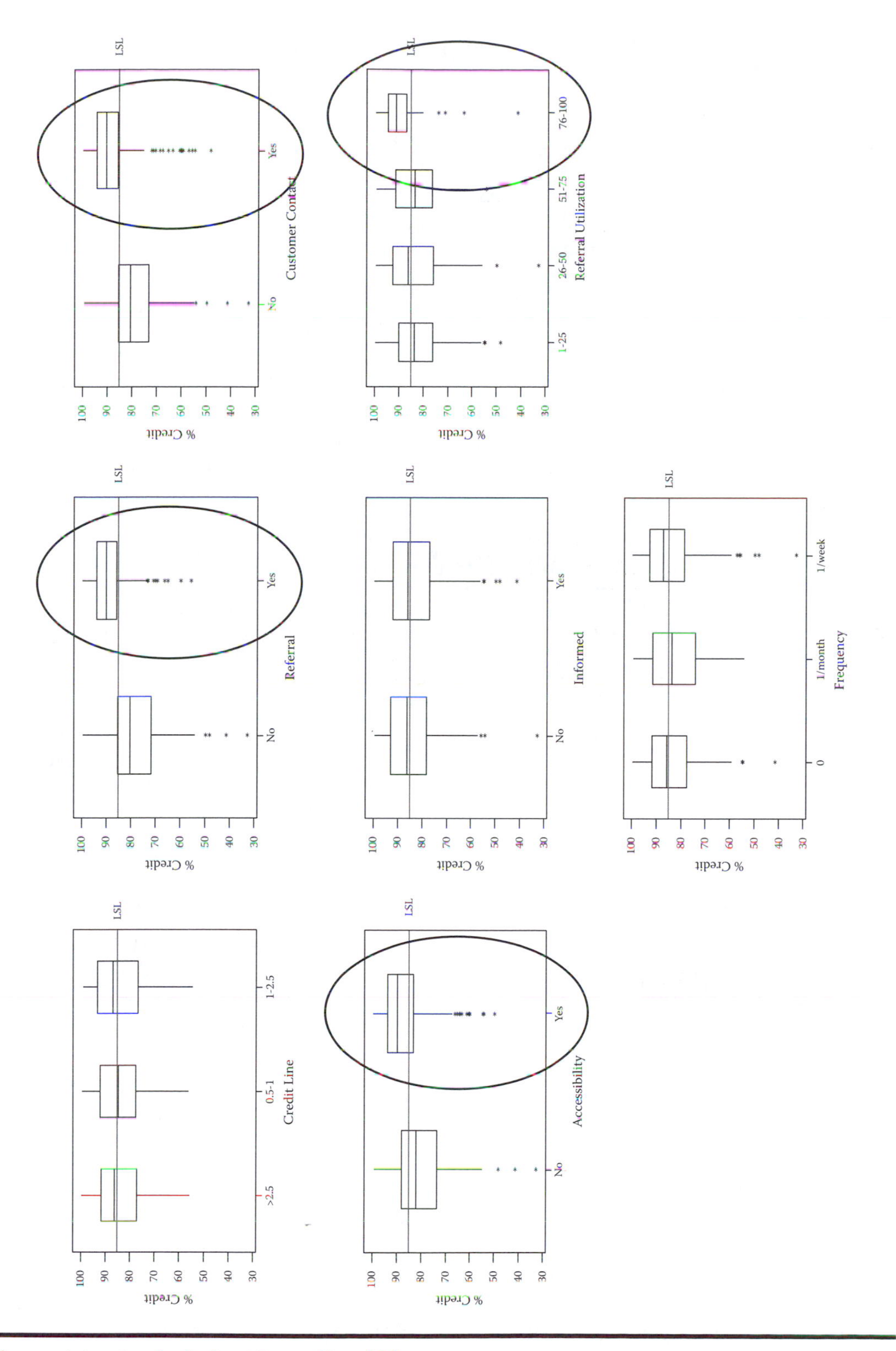

Figure 4.1 Analysis for Unpredictability.

Table 4.2 Statistical Results Indicating Significant Factors

Source	*DF*	*SS*	*F Ratio*	*P*
Credit Line	2	0.138185	0.6110	0.5434
Referral	1	13.811689	122.1369	<.0001[a]
Assigned Contact	1	10.786436	95.3846	<.0001[a]
Accessibility	1	2.336272	20.6597	<.0001[a]
Informed Customer	1	0.037597	0.3325	0.5646
Referral Utilization	3	2.250331	6.6332	0.0002[a]
Contact Frequency	2	0.554119	2.4500	0.0878

Note: DF = degrees of freedom; SS = sum of squares.

[a] Statistically significant.

Table 4.3 Statistical Results of Analysis Indicating Significant Factors

Term	*Estimate*	*Std Error*	*t Ratio*	*P*
Intercept	0.831823	0.005164	161.08	0.0000[a]
Referred	0.040434	0.005311	7.61	<0.0001[a]
Customer Contact	0.030422	0.005351	5.69	<0.0001[a]
Accessibility	0.019040	0.005277	3.61	0.0004[a]

[a] Statistically significant.

The optimal settings for the significant factors are: Referral = Yes, Agent = Yes, and Accessibility = Yes, which give the highest predicted percent credit utilization of 92.16%.

3) Addressing Unpredictability

While the issue of revenue generation was addressed as an unpredictability problem, there were some actions that the company could take. None of the factors caused the customer to borrow more, but there appeared to be factors that could be influencers. The actions they chose are listed below:

While the company did not want to restrict itself to referrals, it started a marketing campaign to solicit referrals from existing customers. This will increase the number of customers with Referral = Yes.

The company assigned an agent to each customer with a credit line above $1 million. This will increase the number of customers with Agent = Yes.

Customer-contact employees were reluctant to give their customers their home numbers. To address accessibility, the company offered to pay these employees' cell phone plans if a) they forwarded their home numbers to their cell phone and b) gave their home number to their customers. This will increase the number of customers with Accessibility = Yes.

To address the frequency of contacts between the agent and the customer, the company instituted a policy where the agent would ask the customer how often and for what purposes they wanted to be contacted. This is a check on whether it is mere accessibility or a certain amount of communication.

A more rigorous solution would include changes to sustain the improved performance. The team set up several monitors:

The company tracked all new referred customers against new nonreferred customers on utilization.

The company tracked utilization by customers with agents against those without agents.

Not all agents took the calling plan subsidy, so the company monitored whether those who did had customers with higher utilization than those who didn't.

Quarterly reviews were set up for ongoing validation of the prediction model using the original database and all the new loans and results since the model was installed.

The company also began modeling using other factors.

What You Can Do Now

As with suboptimality problems, it may not be a benefit to learn how to solve unpredictability problems unless it is part of your job to make predictions. In which case, there are four things you can do to learn more about prediction models and gain expertise in creating them:

- Learn more about mathematical modeling, e.g., design of experiment, Kepner–Tregoe, Shainin.

- Learn using statistical and mathematical software designed for modeling.
- Practice using these tools on processes you use at work and at home (e.g., predict any of your utility usage by month), and, on products or processes you have already created, similar to a postmortem.
- To teach others, review the training chapters (Chapters 11 and 12) and create simulations so that others can learn by doing.

Chapter 5

Personal Reason-Caused Defect

How to Identify Personal Reason-Caused Defects

"I'm not paying for that service!"
"I've decided not to renew with your company."
"I didn't know I had to do that."
"If you aren't going to…, then I'm not…"

This problem can be identified when the person causing the defect does not think it is a defect, but, from a process perspective, it is a defect.

If this type of problem is systemic, a company can typically get information about it from customer surveys. The company will look for trends and common responses, as they provide clues to systemic issues. However, when the situation is personal, a survey is not likely to get at that issue. For example, a customer may not renew a contract because a single incident became a personal reason not to renew.

Some defects caused by personal reasons are due to forgetfulness or lack of knowledge. These kinds of personal reasons cannot be addressed unless you ask the person, especially when it is due to lack of knowledge.

Others are consciously caused. You probably have caused such defects, but not from your perspective. You leave one employment for another. If the company valued you, then they will view it as a defect—you don't.

On a rare occasion, you know you are causing a defect, but believe you are justified in creating it. Have you ever defiantly not done something that should have been done or delayed doing something that should not have been done? These are quid pro quo actions for a defect you received. Asking "why" helps you understand the real process problem (the defect that led to your action) so it can be addressed.

Personal reason defects can occur in all processes requiring people to behave in a predefined manner. For example, a customer not renewing or not paying an invoice, a high-potential employee leaving the company, a plant employee not wearing safety equipment, a new hire not completing required online training, or a frequent traveler not turning in expense reports.

Key Words: accounts receivable (high), employee turnover (high), customer loyalty (low), policy (violation), (non)compliance.

How to Find Root Causes of Personal Reason-Caused Defects

Ask why or why not:

If a customer does not renew or does not pay an invoice, ask why not.
If a high-potential employee is leaving the company, ask why.
If a plant employee is not wearing safety equipment, ask why not.
If an employee does not turn in an expense report, ask why not.

The only way to discover that personal reason may be to ask the person him/herself. We often say that the reason we didn't do this or that is because … yadda, yadda, yadda. The reason is the cause.

How to Address Personal Reason-Caused Defects

Do something about it—or move on—For service companies (which most companies are), learning the personal reasons is the first step in being customer-oriented.

Sometimes the reason identifies another type of root cause, e.g., an error-caused defect. For example, if the reason a customer does not pay an invoice is that she only got one of the two batteries she ordered, you won't know unless you ask. In this case, the root cause is the error.

If the battery had been delivered late, it becomes a delay type of problem. If the battery doesn't work, it could be a suboptimality type of problem, and so on.

The second step sets the tone of customer relationship, as in: Are you willing and able to address the reason? For example, one company may automatically replace product that is damaged during shipping with no questions asked, while another will allocate resources to first deny responsibility, then, if responsible, determine the extent of damage, how much to compensate if liable, and method of compensation (e.g., refund or discount on future purchase).

In the battery example, the company can ship the second battery at the company's expense. That addresses this instant, so the action is a corrective action. It, however, does not address the process problem. Preventive action treats the process as having an error problem. Note that the process problem is not a personal reason defect. If you are willing and able to address the personal reason, then do so. If not, say so and move on. However, know that the defect may continue to occur.

If this is a frequent customer, not taking preventive action results in a repetition of the corrective action—at a cost to the company each time. And, the customer may eventually leave.

If other customers have the same reason because of the same defect, you will not know of the real process problem unless you ask.

Be aware that a personal reason may change. An employee may provide one reason for leaving this year, but changed circumstances may change the reason. For example, low pay may be a reason one year and inflexible hours may be a reason the next year due to the birth of a child. If you do not ask periodically, you will not know of the change.

One key is to understand where personal-reasons-caused defects can occur. The first section of this chapter lists several examples. Use your own experience of personal reasons or when you wanted to created a defect through a personal reason to understand where they can occur.

Personal Reason Example 1

In a regulated industry, a company must take corrective action. During an internal audit, the auditor discovers that there are four corrective actions that were not processed. The auditor asked the individual responsible why this wasn't done.

Table 5.1 Disposition of Personal Reasons

Person	*Reason*	*Disposition*
1	Person still working on it, waiting for decision by manager	Remove as defect
2	Person was unaware that ...	Treat as personal reason root cause; now that the person knows, she said she would follow procedure
3	Person said there was an error on the report; the SOP reference was wrong	Treat as error root cause
4	Person did not believe it was a corrective action issue	Treat as personal reason root cause

1) Identify Personal Reason Defect

Table 5.1 shows the results of step 1. For each person, the reason his/her action resulted in a defect is noted. The disposition of that reason means recognizing the real cause is another type of defect or it is an actual personal reason.

2) Find Personal Reason Root Cause

In the second case, the personal reason was lack of knowledge. Once the person acquired the knowledge, the defects caused by this personal reason should cease. In the fourth case, the personal reason was a disagreement with the auditor's view.

3) Address Personal Reason Root Cause

No analysis is required other than determining (1) whether the reason is due to a defect and if not (2) whether the company was willing and able to address the reason.

Person 1: Not a defect, but could be treated as a delay if timeliness is the issue.

Person 2: Unawareness was resolved by the fact that by asking about the issue removed the lack of awareness. Now that the person knows, she said she would follow procedure.

Person 3: Treat as an error root cause by (1) mistake-proofing and (2) refining standard operating procedure with what to do when errors like that do occur so the process still continues.

Person 4: This was a real personal reason root cause that the person's manager was willing and able to address. First the person was told that this was a legal issue and the company had to comply and it was the person's responsibility. The person said that that was true, but not if it isn't a corrective action item. So, she wanted to know how that is determined.

The person explained to the auditor why he thought it was not a corrective action issue. The auditor explained why it was and provided documentation to clarify their differences. The person responsible understood and said he would change his view and actions.

Note that this personal reason could not be identified without significant work any other way. A survey would have to be precise about this reason, which would be unlikely without knowing ahead of time. However, now that this reason is known, the company can issue a memo or provide training to all appropriate personnel to clarify the issue.

Person 4 filed the report and took action. The manager gave the person the responsibility to develop criteria and a procedure for evaluating whether an item is a corrective action or not. The person did and the process was changed to include both. This was followed by (1) providing training on the change and criteria including test cases, (2) creating a visual display of corrective actions and their status, and (3) retesting on both how to process corrective actions and how to resolve nebulous cases.

Personal Reason Example 2

1) Identify Personal Reason Defect

The owner of a car dealership surveys his customers quarterly. Based on the survey, he implements improvements to the services he provides them. One such change was extending the distance the dealer will travel to provide free shuttle service for owners whose cars are in the service shop. In addition, he receives complaints via email and letters. One complaint was not caught by the surveys, but did lead to improved customer relations.

A customer had dropped off her car for a 10,000-mile checkup. In her hurry, she failed to turn in a coupon for services she had paid for in advance. When she got the credit card bill, she realized her mistake. She returned to the dealership and spoke with the service manager. After a half hour, she left dissatisfied with the response and sent a letter to the owner explaining her situation.

2) Find Personal Reason Root Cause

The owner contacted several customers who sent complaints. Three responses are shown in Table 5.2, including the aforementioned customer as Person 3.

No analysis is needed other than determining (1) whether the reason is due to a defect and, if not, (2) whether the company was willing and able to address the reason.

Person 1: A defect, but could not be addressed.

Person 2: A defect, although no customer dissatisfaction, but could not be addressed.

Person 3: This was a real personal reason root cause that the person's manager was willing and able to address. The owner called the customer, apologized, and accepted the coupon. The customer said she would return to that car dealer for her next service.

3) Address Personal Reason Root Cause

The owner worked with the service manager to develop a procedure where rather than wait for customers to provide any prepaid coupons, they would ask the customer if they had any. Owner added a question to the survey that

Table 5.2 Disposition of Causes by Type of Personal Reason

Person	*Reason*	*Disposition*
1	Person provided neither sufficient detail nor contact information to identify issue and person	Defect, not able to address
2	Person was pleased with the service, but was moving from the area	Not able to address
3	Person said the service manager argued with her and refused to accept the coupon because it wasn't provided at the time of service	Dealer owner decided it was a personal reason he was willing and able to address

related to coupons and continued to review survey results and unsolicited feedback.

What You Can Do Now

Systemic issues resulting in undesirable behaviors can be discovered through anonymous surveys to an appropriate sample of people. General knowledge does not lead to specific solutions. If you have thousands or even hundreds of people, personal issues may not be worth the effort to solicit, but you can accept them. That is what complaints are about. If the number is small or manageable, then it may be worthwhile soliciting them.

However, when the undesirable behavior is personal, the only way to ensure you have the right information is to ask each person individually. Manageable may just mean that your sales force, for example, can get the information rather than by marketing via mass mailings.

For example, trying to keep high-potential employees will most likely require knowledge of each individual's personal reasons. A personal conflict will not be resolved through an anonymous survey. Reasons why customers do not renew or buy more may be for personal reasons, e.g., a single specific event has alienated the customer.

Personal reasons can reveal a defect (systemic or not) and you should then treat it according to the defect type. When the behavior is not due to a defect, then the personal reason is truly the root cause. When the personal reason is unawareness, it may resolve itself just by the asking. Even if it is lack of skill, the asking can quickly lead to resolution.

When it is more complicated than that, then the personal reasons fall into two categories: things you are willing and able to address and those you are not. When you see undesirable behavior resulting in defects, ask the person. There are four things you can do to acquire the habit of identifying personal reasons and expertise in addressing them:

- Learn more about listening skills, asking questions, and negotiation.
- Study tools used to actively listen, interview, negotiate.
- Practice using these tools with people you interact with at work and at home (e.g., when, as a customer, you are dissatisfied, determine first if the issue is unique to you or systemic—ask the vendor; if unique, determine what specific action you want the vendor to take—tell the vendor).
- To teach others, review the training chapters (Chapters 11 and 12) and create simulations so that others can learn by doing.

MULTIPLE UNIT PROCESS IMPROVEMENT

Principle

Recall Lean's five principles:

1. Identify value from the end customer's view
2. Map the value stream
3. Create flow
4. Create pull
5. Seek perfection

In this section, you will use principle 1 to expand your ability to improve processes. For this principle, you will learn a complete and comprehensive definition of value.

The other principle you will use is the fifth principle: seek perfection. There are three applications of this principle. This section covers two applications: chronic process problems and designing processes. The third application is covered in Section III.

Leveling

A complementary technique to jidoka is *heijunka*. Heijunka means leveling. In Lean, it applies to production. It is a way of creating flow by having constant work rates. Leveling of the work load across consecutive operations prevents buildup and stop-and-go production. It is sometimes viewed as the slow-and-steady tortoise rather than the sprint-and-nap hare.

The philosophy of heijunka applies to multiple unit process improvement by leveling to reduced levels the defect rates by type of defect. To most effectively apply heijunka in this context of solving process problems, you will look at multiple units from a process. This does not mean that the process can or should process more than one unit at a time. Multiple unit process improvement just means that you will look at a recent history of process problems to determine how to proceed. It is applicable when you have not made improvements to your process in a long time, if ever.

It would not be surprising if every company is in this situation with some of their processes. Learn first by knowing how to tackle single unit problems. Gain experience with them and then move to chronic problems.

Chronic Problems

Eventually you will have projects to improve processes that are too big for one person to handle. What you learned in the chapters on the five types of process problems is how to solve a single occurrence of a particular defect.

You could do this for all processes—wait for a defect to occur on a single unit and address it. Or, you can be proactive by identifying a potential defect and addressing its potential cause, e.g., mistake-proofing, removing NVA (nonvalue-added) actions, checking optimality, validating prediction factors and models, asking people.

Chronic problems will occur when the defects have not been addressed for awhile. People have been living with the defects, but now they want and have support to address them. This process improvement is reactive. People want to know they have made an improvement.

This section addresses chronic problems. You already have processes that are running and producing defects. Working on a single unit for a single defect is fine for helping you acquire a continuous improvement perspective and habit, and learn basic concepts, tools, and procedures for process improvement. To increase your ability to improve processes and accelerate the benefits gained from such results, you move to improving processes by addressing more than one defective unit and addressing more than one type of defect.

Rather than having one person working on improving a small process, you might have multiple people working as a team on a larger process with more than one type of problem.

The difference now is that you will need to establish a baseline performance. In the Introduction, you learned that process performance is just the percent of good products or services produced by the process for a period of time. To establish a baseline, you will need to define certain things and collect data.

After you make improvements, you will check that you have improved from the baseline. You need to show that the percent good products or services for a subsequent period of time is better than the baseline percent good.

In this section, you will learn to improve processes that have not been improved for some time. The defect rates for various defects are typically unknown. Rather than starting with a single unit and a single defect, you will look at the defect rates for several defect types across a recent history. Then you can prioritize and begin addressing the most pressing problem, resulting in a leveling of defect rates *to zero*.

Critical Thinking

Because you are focusing on multiple units and possibly various defect types, you will use an expanding version of the three-step procedure you already learned. However, to enhance your ability to solve process problems because you have a deeper understanding, the procedure will be based on critical thinking. Critical thinking relies on information. Types of information include facts, evidence, and subject matter expertise.

Informative Decision Making

Lean Six Sigma is an information-based, problem-solving process. Whether you are eliminating waste or reducing variation, you need information for the decisions you are required to make throughout the problem-solving process. From selecting and prioritizing problems and projects to finding root causes to finding and implementing solutions, you should make decisions based on information.

For this reason, the critical thinking embedded in the different methodologies (e.g., three-step procedure, DMAIC (define, measure, analyze, improve, control) for existing processes, DFSS (design for Six Sigma) for new processes, Lean) you use to improve your processes are structured to recognize the need for information. Critical thinking has two levels. The first level consists of questions. Every question has the format: “How do I know …?” to make clear the need for information.

The second level identifies the information you need to answer the questions. It identifies the information needed, so "you know… ." In other words, a question of the type: "How do I know (this)?" is answered by: "Because I know (that and that)."

Critical Thinking Questions

Keeping with the theme of this book to simplify, there are four phases to solving chronic process problems. Each phase asks and answers a critical thinking question. Objective problem solving requires that your answers are based on information. The four critical thinking questions and information needed to answer them are specified:

1. How do I know I have a process performance problem?
   - I know the customer requirements.
   - I know how often I have met the customer requirements (current or baseline process performance).
   - I know how often I would like to meet the customer requirements (desired process performance).
   - I know the baseline performance is less than the desired performance.
   - I know it is worth closing the gap.
2. How do I know the root cause?
   - I know the type of process problem.
   - I know which factor(s) changed the process performance.
3. How do I know the proposed solution works?
   - I know that, with the solution, performance improves and, without it, performance does not improve.
   - I know I have mistake-proofed the change.
   - I know how to sustain the improved performance.
4. How do I know when I can improve again?
   - I know when a signal indicates a change for the better and how to respond to such signals.

Continuous improvement is cycling through increasing levels of improvement or from current to desired performance. Figure II.1 shows how the critical thinking questions are related to the continuous improvement cycle.

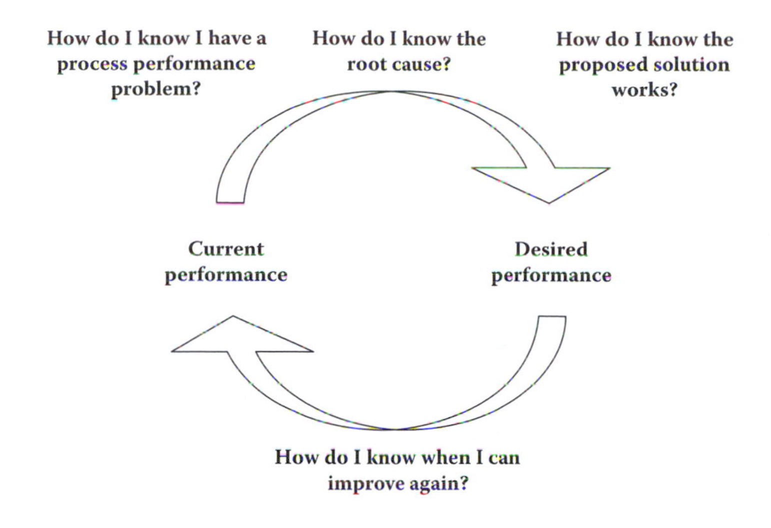

Figure II.1 Critical Thinking Questions and the Continuous Improvement Cycle.

Information

You know the answers to the critical thinking questions when you have specific information. Information can come from many sources including facts, empirical data, theory, scientific laws, evidence, and subject matter expertise. Table II.1 shows what specific information you need to answer each question.

Tools

Tools (e.g., templates, analyses, procedures, software) are ways of generating information. To become more proficient with tools, associate tools with information. Learn what information you need and then what tools will help you get it. This book does not focus on tools. For the most part, tools are not even mentioned except generally by describing templates, providing procedures, and recommending ways to organize your searches and findings. The exception to this has been for suboptimal and unpredictable causes of defects. In these two cases, statistics was the general tool. The reason is that problem solving can be done without tools.

There are a variety of tools, but focusing on tools typically creates a barrier to deeper understanding. A practitioner's first questions should not be "what tool do I use?" Rather, the first questions should be either: "What do I need to accomplish (or deliver)?" or "What do I need to know?"

One tool to facilitate your understanding and application of critical thinking to process improvement—whether individually, as a mentor, or on a team—is

Table II.1 Critical Thinking Questions and Information Needed to Answer Each Question

Critical Thinking Question	*Specific Information*	*Source*
1. How do I know I have a process performance problem?	a. Requirements b. Baseline performance c. Desired performance d. Business impact (value of closing gap)	a. Customer/market b. Process c. Business d. Financial analysis
2. How do I know the root cause?	a. Type of defect b. Causal relationship by type c. Data on relationship between cause and process performance	a. Requirement, defect b. Process c. Dependent on defect type—output units
3. How do I know the proposed solution works?	a. Proposed process change b. Evidence that change closes performance gap c. Evidence that improved performance is sustainable	a. Process b. Piloted solution c. Mistake-proof or piloted controls
4. How do I know when I can improve again?	a. Signal indicating unplanned improved process b. Procedure	a. Controls b. Self

a critical thinking template. Such a template would have two more columns (see Table II.2). The first column lists the critical thinking question. The second column contains the list of information needed. The third column lists the sources of that information. The fourth column lists possible tools (e.g., other procedures, documents, forms based on definitions, guide lists) embedded or linked to the table cell to record and organize the information you collect. You also might have a fifth column labeled: "How did I answer the question?"

Example

Suppose you are just starting a project on a chronic problem. You are trying to answer the first critical thinking question: "How do I know I have

Table II.2 Template for Recording Application of Critical Thinking to Chronic Process Problems

Critical Thinking Question	*Specific Information*	*Source*	*Tools*	*How Question Was Answered*
1. How do I know I have a process performance problem?	a. Requirements b. Baseline performance c. Desired performance d. Value of closing gap	a. Customer/market b. Process c. Business d. Financial analysis		
2. How do I know the root cause?	a. Type of defect b. Causal relationship by type c. Data on relationship between cause and process performance	a. Requirement, defect b. Process c. Dependent on defect type—output units		
3. How do I know the proposed solution works?	a. Proposed process change b. Evidence that change closes performance gap c. Evidence that improved performance is sustainable	a. Process b. Piloted solution c. Mistake-proof or piloted controls		
4. How do I know when I can improve again?	a. Signal indicating unplanned improved process b. Procedure	a. Controls b. Self		

a process performance problem?" which is listed in the first column and first row.

The second column (Information) states that to know that you have a problem requires that you know five things:

a. Requirements
b. How often you have met customer requirements (baseline performance)
c. How often you need or want to meet customer requirements (desired performance)
d. Whether the baseline performance is worse than desired performance
e. Whether it is worth closing the gap between baseline and desired performance

The order guides you from the first thing you need to know to the answer to the critical thinking question. You cannot calculate baseline performance without knowing the requirements. You cannot know if there is a gap until you have both baseline and desired performances, and, so on.

If your improvement initiative requires specific forms to record your answers, then embed or link them in the fifth column: "How did I answer the question?"

Summary

Each of the chapters (6 through 9) addresses one of the critical thinking questions. Chapter 10 addresses designing processes with its over-arching critical thinking question. Each chapter ends with an example and summary of the information needed for that critical thinking question.

Chapter 6

How Do I Know I Have a Process Performance Problem?

Because you know you have a gap that is worth closing, and a problem is a gap worth closing.

In Section I, you considered only single output units that had a defect. We assumed all defects were worth addressing for the purpose of teaching the five types of problems, how to identify their root causes, and how to find solutions addressing the root causes. However, in practice, not all gaps are worth closing.

Solving chronic process problems involves more than just looking at single units. You need to look at process performance. Recall that process performance is the percent good for a period of time. How do you know if you have a process performance problem? It's simple. You know you have a problem when you know it is worth closing the gap between your current process performance and your desired process performance.

In this chapter, you will learn how to acquire information to know if you have a problem. That information builds as follows:

- Gap between current and desired process performance:
  - Current process performance:
    - Requirement
  - Desired process performance
- The value of closing the gap:
  - Consequence or benefit

To know whether there is gap, you need to know both current and desired performance. To know current performance, you need to know the requirements. To know the value of closing the gap, you need to know the consequence of not closing the gap or the benefit of closing it.

Requirement: Identifying Value

Before you can calculate any process performance, you have to know whether the output unit is good or bad. The first Lean principle is to identify value from the end customer's viewpoint. In Section I, value was defined as a product or service being defect-free. Being defect-free would be of value to a customer. You learned how this is applied to each single unit of product or service.

But, a defect-free product that the customer did not order or want is no longer of value to him. Without knowing what the customer wants or needs, you cannot determine whether that product or service is of value, even if it is defect-free. In other words, you need to know more specifically what counts as a defect. That the product or service functions is not enough.

In the first section, you were shown how to identify a process problem by the type of defect. You were taught there were defects caused by delays, errors, suboptimality, personal reasons, and unpredictability. However, you were not told specifically why something was a defect. In most cases, you could imagine why. For example, in a delay-caused defect, you could imagine that the person wanted it before he or she got it. However, you may not know at what point it would be considered late.

Think about a product you have purchased. You identified the product that you wanted. You looked at it to see that it had the features you wanted. For example, you wanted a car. One feature of the car that is important to you is its color. The colors available at one dealer were forest green, fire red, midnight black, and sky blue. You wanted sky blue.

The same thing occurs with services. You decide what service you want and then see whether it is accomplished in the way you want. At work, you go out for lunch. One feature of lunch might be that it be delivered quickly. You want to be done eating in 20 minutes, not 30 or 45 minutes.

What do these two descriptions have in common?

- There is a thing you want.
- There is a feature of the thing that is important to you.

- The feature comes in several choices.
- Only some of the choices are acceptable.

From these general descriptions and examples, you can identify four components of a requirement:

1. What is the product/service?
2. What is the feature or characteristic (fit, form, function) of the product/service that could be defective?
3. What are the choices for (how do you measure) the characteristic on a single unit of product/service?
4. Which choice(s) do you want (what are the specifications)?

The characteristic can be how a product functions, dimensions of a product, properties of product (e.g., softness, durability, flexibility, porosity), aspects of service including timeliness, quantity, delivery method and manner, delivery location, convenience. These characteristics are sometimes referred to collectively as "fit, form, and function."

Because the product or service is what comes out of the process, let's call it the *output unit*. It is typically easy to know what the output unit is of a process. Similarly, identifying the characteristic is usually easy to understand. The measure is more difficult.

Think about it this way. The purpose of measurement is to make distinctions. When you measure something, there are different results you can have. The different results are your choices. They help you distinguish things, between the fire-red car and the sky-blue car, for example.

The choices help you compare. If you want a blue car, then the ability to "measure" (you may do the measuring of color with your eyes) is the ability to distinguish between the sky-blue car that you want and the cars of other colors that you do not want.

The number of choices could be as few as two: good or bad, blue or not blue, on time or not on time. You can give the appearance of being quantitative by labeling these choices 0 and 1, for example. Choice of color might appear to be infinite, but the car manufacturer may actually limit the choices to only five colors. In all of these cases, the measure is simply the list of possible choices.

When the choices are infinite and quantitative, such as dimensions, you can use mathematical notation to make the list. For example, >0 or $\{1, 2, 3, \ldots\}$. Because you cannot measure to whatever degree you want, you can list what is called the resolution of the measurement or the smallest amount measurable. For example, if you measure to the nearest tenth of an inch, you can write the list of choices for the length of table as >0.0" or just simply note the resolution, e.g., 1° for temperature, 0.1 seconds for speed of a robotic arm, $100 for general ledger expenses.

The choices that are acceptable define a *good product* or *service.* For example, you might define a good invoice as one that has "the correct name, address, items purchased, quantities purchased, and the total amount including taxes, shipping, and handling charges." This would be the definition of good. Conversely, you might choose instead to define *bad* by saying an invoice is bad when any of the following is incorrect: name, address, list of items purchased, quantities purchased, taxes, shipping, handling, or totals.

When it comes to quantitative metrics, the definition of good is typically a range of possible values. For example, a table manufacturer builds 6-foot tables. The manufacturer knows the tables won't be exactly 6-feet long, so it tolerates 1/32 in. error more or less. The range would be from 5 ft 11 31/32 in. to 6 ft 1/32 in. This type of requirement is called a specification.

There only three cases possible for requirements and they determine the types of specifications and whether you have a target (or nominal):

1. More is better: A lower specification limit (LSL) is needed, but no target. For example, to increase revenues, a sales force is tasked with bringing in more leads. A lower specification limit represents the minimum number of leads.
2. Less is better: An upper specification limit (USL) is needed, but no target. Time to process is the typical less-is-better example—a pizza delivered after 30 minutes is a defect. A maximum amount defines the USL, but no target is necessary.
3. A specific value is better: Both LSL and USL are needed and the specific value is the target. For example, a table manufacturer states the length of the table (specific value) as 72 in., but permits a tolerance of 1/32 in. So, LSL = 71 31/32 in., Target (nominal) = 72 in., and USL = 72 1/32 in.

Requirements define when a product or service is defect-free based on what the customer wants or needs. When the feature of the product or

Table 6.1 Examples of Requirements by Defect Type

	Product/ Service	*Characteristic*	*Measure (Choices)*	*Specifications (Good)*
Error	Two-battery order	Amount of batteries received	Number of batteries: 0, 1, 2, …	Two (anything else is a defect)
Delay	Two-battery order	Delivery date of second battery	Number of days after promised date: 0, 1, 2, …	0 (anything else is a defect)
Suboptimality	Two-battery order	Battery size	Whether it fits in camera's battery slot: yes/no	Fits or yes (No is a defect)
Unpredictability	Two-battery order	Inventory	Number of batteries exceeding average monthly shipped: 1%, 2%, 3%, …	15% or more (less than 15% is a defect)
Personal Reason	Two-battery order	Payment	Number of days after shipped payment received: 0, 1, 2, …	Within 30 days of shipment date (more than 30 is a defect)

service fails to meet the specifications, there is a defect. Table 6.1 gives examples for each type of process defect.

The first three examples are from the customer's perspective while the last two are from the supplier's. The first three defects clearly cause the customer to be unhappy. The last two also could result in an unhappy customer. If unpredictability results in a late delivery or out-of-stock condition, the customer would be unhappy. If the personal reason of not paying because of lateness or other defects causes the supplier to threaten or initiate legal action against the customer, the customer might be unhappy. Unpredictability and personal reason defects cause the supplier to be unhappy because they have profitability consequences for the supplier.

Consider unpredictability. The vendor might budget and take actions based on the assumption that payments are made within 60 days of shipping. Some action might include making other purchases or hiring additional personnel. Defects might reduce cash flow costing more money or reducing margins, and reducing profits.

The personal reason might be a result of the first type of defect. The customer gets only one battery and immediately calls the credit card company disputing the charge. This delays payment to the vendor. The supplier would need to address this defect.

A true personal reason might be the customer was defrauding the vendor using a stolen credit card.

Current (Baseline) Process Performance

Now that you know the requirements, you can determine the current process performance, or, baseline performance. *You will need to do this for each requirement.*

Process performance is *the percent good for a specified period.* This creates another measure. This measure is not of the characteristic of a single unit. It is a measure across multiple units. It tells you how often you met customer requirements. Because a process is supposed to repetitively do the same thing and get the same result, process performance is measured across multiple units. Table 6.2 shows examples of the difference between a measure for a requirement and a measure for process performance.

Table 6.2 Examples of Output Unit Requirement Measures and Process Performance Measures

General Requirement	*Output Unit Characteristic Measure*	*Specification(s) on single Output Unit*	*Process Performance Measure*
Quality, e.g., length	Cm for diameter of a tube	LSL=1.9 cm, Target=2.0 cm, USL=2.1 cm	% meeting length specification
Time	Seconds from online order completed to receipt of online confirmation	USL=4 seconds	% meeting time specification
Cost	Cost per unit	USL=$0.71	% meeting cost specification

The measure for a requirement is the measure of the output unit's characteristic. The specific requirement are specifications for a single output unit. Process performance is the percent meeting the specifications, or, percent good.

To determine current process performance, take these steps:

1. Select a recent period of time.
2. Select all the output units that were processed during that period or randomly select a sample from that period.
3. Determine the total number of output units you selected.
4. For each output unit, (1) look at the characteristic, (2) measure that characteristic, and (3) determine whether it is good or bad by comparing to the specifications.
5. Count the number of output units that were determined to be good.
6. The simplest calculation[1] of process performance is the percent good: 100% x (number of good output units)/(total number of output units).

This seems straightforward.

However, often data collection is not easy, especially if you will not look at all the units in the period selected. Statistics can help you take a sample that is useful.[2]

Random Sampling

There are other ways of calculating baseline performance that might be more accurate, especially with small sample sizes. Consider this situation. If you collect data on only 10 units, then your baseline performance as a percent good can only be one of the following: 0%, 10%, 20%, 30%, 40%, 50%, 60%, 70%, …, 100%. If it is important to know whether it is 70% or 73%, then you need to either collect data on more units or use statistical modeling.

Statistical modeling requires an assumption about how the population from which the data came from behaves. For example, if the population has a certain distribution, you can use that fact or assumption to calculate a more precise estimate of the percent good. A distribution means that you know or assume the frequency of each possible choice. For example, a fair die has a known distribution: Each of the six possible choices occurs equally often. More sophisticated calculations of percent good for quantitative measures are possible when the distribution follows particular probability models.

Process Capability

Often, the percent good is calculated under a variety of assumed conditions. These different calculations get at not only the current performance of the process, but also the potentially best performance of which the process is capable. Regardless of whether it is actual or potential or current or future or hypothetical, all of these capability calculations are simply estimating the percent good. In other words, process capability simply means how often did or do you expect the process to meet customer requirements (under the assumed conditions in which the calculations apply).

Estimating the potential process performance is useful in the development and design phase of processes. You want to know how good the performance can be before you spend time and money creating and implementing it. You don't want to install a process that is so poor that it's a losing endeavor.

But, once the process is running and producing output units, the current performance simply is what percentage actually met the specifications for a recent time period. That's what is produced and what the customer gets (or will get).

Desired Process Performance

A problem is simply a shortfall between what is currently happening and what one wants to happen. With respect to process performance, a problem is a shortfall between current performance and desired performance. If the current performance equals or exceeds desired performance, then there is no process performance problem. At least, there is no problem with respect to the customer requirement for which performance was calculated.

Therefore, it is necessary to determine the desired process performance with respect to the customer requirement of interest to determine whether there is a problem and to do this for each requirement.

There are several reasons why it is more efficient to wait until this point in the problem-solving procedure to determine whether there is a problem and how to define it well:

- Getting customer requirements is critical. If you do not know what they are, then the problems you think you have are obvious ones: the first two defect types of errors and delays. If they are not obvious, then obtain the requirements to ensure you solve the right problem.
- Getting customer requirements is not always an easy task. Do that before you start defining a problem so you won't have to rework it.
- Getting the data to determine whether there is a gap between current and desired performance also is not always an easy task. Do that before you start defining a problem, and then have to rework it.

However, it is not the customer who determines the desired process performance. Everyone already knows what percent good the customer wants—100%.

Let's distinguish between what ultimately everyone wants and what is a feasibly desired process performance. The desired process performance should be a project goal based on several reasons:

- The technology does not exist for 100% good given the specifications.
- The characteristic is not what the business is competing on.
- The characteristic is not critical to the market.
- The specifications vary by customer or customer segment.
- The best competitor is not at 100%.

Whatever the reasons, it is the business's responsibility to determine the desired process performance for the purpose of problem solving.

Even if you are practicing continuous improvement, you can set the desired process performance continually between your current and 100%. This follows the fifth Lean principle: seek perfection.

The Gap

Now that you have both the current and desired process performance, you know whether there is a gap.

Typically when trying to improve process performance through project work on multiple-unit results, the project team is asked or required to create a charter. This is especially true when a team is doing the problem solving.

Three common elements of a charter are the problem statement, the goal statement, and the business impact statement. To bring clarity to these issues, we recommend replacing the problem statement with a baseline statement. In other words, a charter should have a baseline process performance statement, a desired process performance statement, and a business impact statement.

To write a clear, unambiguous, and complete baseline statement, you need to include the components of the customer requirements and the components of the current process performance. Generically, a baseline statement on process performance can be stated as follows:

> Baseline Statement: At *where* from *when* to *when*, *how much* (percent good/bad) of *characteristic definition* of *product/service* met (did not meet) specifications.

The first half of the statement uses the definition of process performance with the additional component of *where* to help identify the specific process to which the statement applies:

- *Where* is the geographical location.
- *When:* Because process performance is the percent good during a period of time, the baseline statement must state the times of the first datum and last datum used to calculate the percent good.
- *How much* is the quantification of the performance. The simplest and easiest quantification for everyone to understand is percent. The *percent* can be either the percent good or percent bad. But whichever it is, it must be consistent with the goal statement percentage.

The second half of the statement consists of the four parts of the definition of a customer requirement:

- The *characteristic definition* is the characteristic.
- The *product/service* is to the output unit the customer gets.
- The *specifications* are what defines good.
- The *measure* can be stated in the characteristic definition or the specifications.

Let's look at some examples and see if you can identify all seven elements of a baseline statement.

Example 1

At the Finance Department in the Plainville offices, from January 1 to February 4, 2011, 13% of cycle times from first interview to start date for filling position did not meet HR specifications of 90 days.

Where: Finance Department of the Plainville office
When: January 1 to February 4, 2011
How much: 13%
Product/Service: Fill position
Characteristic: Cycle time
Measure: From first interview to start date in days
Specification: 90 days

Consider this example. Identify the seven elements of the baseline process performance statement before checking your answers.

Example 2

At line 2 at the plant in Plainville, from January 1 to February 4, 2011, 13% of vial content weight of Dimonex did not meet customer specifications of 13 ± 0.5 gm.

Where: Line 2 at the Plainville plant
When: January 1 to February 4, 2011
How much: 13%
Product/Service: Dimonex
Characteristic: Vial content weight
Measure: grams, in 0.1 gm
Specification: 13 ± 0.5 gm

Typically, the difficult part is ensuring all the customer requirement parts are included. These two examples are easy to follow and noting that all elements of a complete baseline statement are included. The reason for this is that the measure (weight in grams, time in days) is quantitative. See if you can find all seven elements in the next example.

Example 3

In the Northeast market, from January 19, 2012 to April 4, 2012, 61% of Nu-Loan approved sales proposals were accepted by the customer.

Is this your answer?
Where: Northeast market
When: From January 19, 2012 to April 4, 2012
Percent good: 61%

Next, the customer requirement parts:

Product/Service: Nu-loan approved sales proposal

Characteristic: Acceptance
Measure of characteristic: (Qualitative) choices are either accepted or not accepted
Specifications: Accepted

Although the statement is simpler, identifying all the parts is harder. This is because, for qualitative measures, people seem to have more difficulty in understanding them as measures and, therefore, understanding the specification. This case seems too simple. The characteristic is acceptance, it has two choices and only one choice is good. That good choice is the specification.

For this reason, it is better to list the seven elements and identify what they are for your case. If you then want to write a statement, go ahead. A sentence is not necessary, just the seven elements.

A desired process performance or goal statement should be connected to the baseline statement so it only needs to state the desired performance and when it is to be achieved:

Goal Statement: By *when*, *extent of characteristic definition of product/service* should meet client specifications.
Example: By October 31, 2012, 80% of Nu-loan approved sales proposals should be accepted by customers.

You have a problem if the baseline good/bad performance is less/greater than the desired good/bad performance.

Worthiness of Closing the Gap

Is the problem worth addressing?—You can't know that until you know what the consequence is of poor current performance or benefit is of better future performance. To make this connection, you need the business impact statement.

Process improvement is consistent with quality philosophies (whether it is Lean, Six Sigma, TQM, quality award, or some other philosophy). All such philosophies are founded on a cause-effect tenet: Increased revenue comes from increased customer satisfaction. This tenet helps establish the following relationship between the three charter statements' baseline, goal, and business impact:

Cause (Baseline Statement) → Effect (Business Impact Loss)
Cause (Goal Statement) → Effect (Business Impact Gain)

The business impact can now be written in one of two ways depending on whether it is the effect of the baseline statement or the effect of the goal statement:

Business Impact Statement for Baseline Statement: The business has lost *extent* during *period* as a consequence of our current process performance.

Example: The commercial loan group had lost sales of $2.1 million between January 19, 2012 and April 20, 2012, as a result of unaccepted Nu-loan approved sales proposals.

Business Impact Statement for Goal Statement: The business will gain *extent* during *period* as a consequence of meeting the desired process performance.

Example: The commercial loan group will gain an average of $654,098 in sales over the same period of three months if it meets the goal of 80% of accepted Nu-loan approved sales proposals.

Prioritization

There are two situations where you will prioritize and not address all defects. You may decide to not close the gap on all defects. It could be some gaps are not worth closing. It could be that you want to get some improvement and benefit as soon as possible.

One case is when you determine there is a process performance gap on several requirements. The other case is when even for one requirement there are several causes for the process performance gap. For example, not meeting a timeliness requirement could due to delays in nonvalue-added (NVA) actions and business value-added (BVA) actions.

If you are not going to address all requirements or all causes, then prioritize. Prioritization can be done by criticality, cost of rework or scrap or both, defect rate, and consequence if defects are not addressed. One way to document information you use to prioritize is to add another column to a requirement table, e.g., as shown in Table 6.3. Prioritization is another difference between single-unit process performance and multiple-unit process performance. You also can show the prioritization graphically with a Pareto chart (a bar graph that displays variances by the number of their occurrences).

Table 6.3 Requirement Template with Consequence Column for Prioritizing

Requirement	*Product/ Service*	*Characteristic*	*Measure (Choices)*	*Specifications (Good)*	*Relative Ranking*	*Consequence Of Defect*	
						Single	*Multiple*

Summary

Below is the information you need to answer the critical thinking question of how to know you have a process performance problem in the order in which you would collect the information:

- Requirements:
  - Product or service (output unit)
  - Characteristic of interest of the product or service including an operational definition
  - Measurement choices for the characteristic
  - Specifications that define when the product or service is good for this characteristic
- Gap between actual and desired:
  - Actual (current, baseline) process performance
  - Desired process performance
- Value of addressing gap:
  - Consequence of not closing the gap, or
  - Benefit of closing the gap

Endnotes

1. Calculating percent good is straightforward—number of good opportunities divided by total number of opportunities times 100. When you have quantitative requirements, there are statistical analyses that can be used to estimate the percentage based on probability theory. The most common analysis of this kind is based on the normal distribution (sometimes called the bell curve). Figure 6.1 illustrates how it is used.

   The vertical lines are the specifications (LSL = lower specification limit, USL = upper specification limit). Any result within the limits is good.

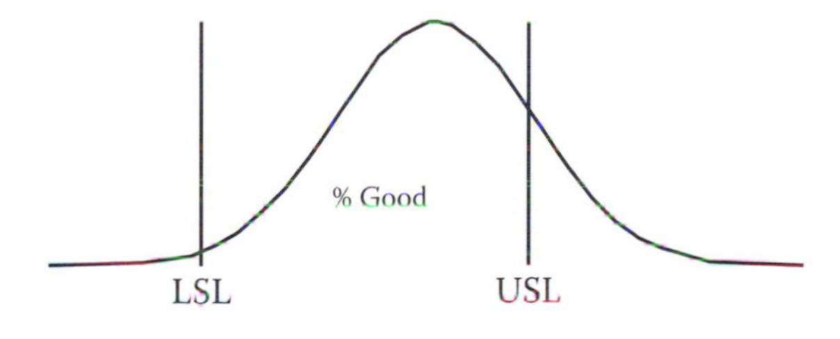

Figure 6.1 Calculating percent good using the normal probability distribution.

The curve represents 100% of the results. The percent of the curve that is between the limits equals the percent good.

2. Data collection for current percent good is simple in concept. You select a time period or a number of opportunities and collect the data. If you are sampling, then the concept becomes a little bit more complex as you want to make sure that your sample is representative of all the opportunities during the period you will consider for your baseline. Various random sampling techniques can be used. The simplest is simple random sampling. Number all the opportunities you could select, use a random number generator to select random numbers, and then identify those opportunities by the number selected.

Chapter 7

How to Know the Root Cause

You know the root cause because you know which causal relationship improves process performance with respect to the requirement.

In Section I, you learned that identifying the type of problem by its defect type reduces significantly the search for root causes. The efficiency occurs not only because you have fewer places in the process to search and you have fewer things that you are searching for, but because, in some cases, there is no analysis or it is greatly simplified.

You can use the type of problem to gain efficiency because it is based on the various causal relationships specific to processes. A causal relationship is simply the relationship between an action in the process and process performance with respect to a requirement.

The good news is that you already know how to find root causes. You learned that in Section I. What is different now that you are trying to improve process performance? What is different when looking at multiple output units rather than one?

Actually, not much.

There are three possible additional things you might do, but the way you find root causes does not change. If you are addressing more than one requirement, then you will have more than one problem type to address. You also will have to calculate process performance for each requirement. Even for the same problem type (e.g., delays, with multiple units), you might have different causes for the delays. So, you will have to search more, but in exactly the same way you already learned.

You will need the following information for each requirement to answer the second critical thinking information: How do I know the root cause?:

- Problem type (for efficient searches of causal relationships)
- Causal relationships by problem type
- Evidence confirming the causal relationship to process performance

Problem Type: Classifying Potential Problems

The first step of the three-step procedure discussed in Section I was to classify the defect type. You will continue doing that, but now you will classify potential defects for each requirement according to the type of defect that occurs when the requirements are not met. After you have gathered the requirements, then classify each by its defect type. In most cases, this should be straightforward. In cases where you are not sure, you may choose more than one classification, but never more than two.

Causal Relationships by Problem Type

Now that you know what type of defect each customer requirement will produce, you can use the second step of the three-step procedure discussed in Section I. For each type of problem, review Chapters 1–5 in Section I to remind yourself of the causal relationships.

Below is a summary of these causal relationships:

- Delays cause a time (e.g., by 10 a.m. or not more than 10 min.) requirement not to be met (Chapter 1)
- Errors cause a correctness (e.g., correct location or amount or size) requirement not to be met (Chapter 2)
- Suboptimality causes a functionality, fitness, or form (e.g., switch does not work, two parts do not fit, or the siding is warped) requirement not to be met (Chapter 3)
- Unpredictability causes an expectation (forecast) requirement not to be met (Chapter 4)
- Personal reasons cause noncompliance that cause any of the above requirements not to be met (Chapter 5)

Use the key words listed in the beginning of each chapter of Section I to help you identify the problem type and understand the general causal relationship.

Evidence Confirming the Relationship between Cause and Process Performance

You can save time by collecting data on root causes when you collect data on percent good to determine current process performance. In Six Sigma language the causes are denoted by X and the effects are denoted by Y.[1]

In many cases, you won't be able or it won't make sense to collect information on the root causes except for when you collect data on the output units. Consider the following situation. You collect information concerning the last 25 surgeries at your hospital to see whether they complied with clinical pathways (standard practices) and find two that did not. You would not look at the next 25 surgeries to find the reasons why these two did not comply.

You may look at both compliant and noncompliant surgeries to find what differences could explain the two that were noncompliant. You could ask the doctors and nurses after the surgeries, but their memory may not be accurate or complete. It makes sense, then, to get the information about the two surgeries during the time they occurred.

There are two types of information you need about each unit:

1. Information to determine whether it is good or bad[2]
2. Information on possible root causes of the bad output units

It should not be surprising that this information varies according to the type of root cause.

Delay

Review Chapter 1 (Delay-Caused Defects). Delay means that something is being waited for. Trace "what is being waited for" upstream all the way to where it originated. During the trace, gather the following information:

- What the wait is for, e.g., for the truck to arrive, for the people in front of me to be served, for the clerk to return from the back room with the item

- When and/or where in the process the delay occurred
- Duration of the delay

Error

Review Chapter 2 (Error-Caused Defects). As with delays, trace the defect upstream from where it was discovered to the first place it could have originated, noting along that path all the places it could have occurred:

- What the error was, e.g., misspelling, incorrect amount, wrong color?
- Where and/or when in the process the error was noted?
- How frequently the error occurred at that point?

You know what the defect is, where and when it was noted, and how frequently it occurred. But, you may not know where the error that caused the defect occurred, only where the defect was discovered.

Suboptimal Setting

Review Chapter 3 (Suboptimality-Caused Defects). For suboptimal setting cases, you want to identify controllable factors as potential root causes. A factor is controllable if you are willing and able to discard choices for that factor that result in undesirable performance. This implies that these factors can vary on their possible choices. These choices are the possible settings for the factor. Some of these choices will result in the greatest percent good. These choices are the optimal settings for the factor. They become specifications for the root causes.

If a factor is not controllable, then even if it is a root cause, you cannot use that factor to improve performance because you are not willing or not able (or both) to set its optimal settings. (You could try to mitigate the effect of that factor by designing something that blocks its effects or influence. For example, hot houses are a result of mitigating the uncontrollable factor of weather or climate.)

Initially, limit yourself to controllable factors as potential root causes.

For each factor that could potentially be a root cause, provide a complete description:

- A label or name (often the labels are X1, X2, etc.)
- A definition

- The metric (list of possible choices)
- For quantitative factors, the resolution (smallest amount measurable)

This root cause can require one, two, or three statistical analyses on three or more data sets. The first analysis is to establish correlation; the second, to establish causation; and the third, to optimize. Using historical data, you determine which factors are correlated with percent good. Generating new data, you determine which of those correlated factors are actually causally related to percent good. Once you know which factors are root causes, you again generate new data to show what settings or choices (values) of the factors are optimal. In the third analysis, you develop specifications for the root causes.

These three analyses are typically used when you have historical data and somewhat of an idea on which of many factors might be root causes. To reduce the large number of factors without too much expense, you use historical data. However, because correlation does not mean causation and causation requires controlled studies (called designed experiments or design of experiment (DOE)), you need to collect additional data on those correlated factors to show causation.

In some cases, you can skip correlation because causation cannot occur without correlation. If you do not have historical data, or subject-matter experts believe the root causes are among a few factors, you may choose to skip the historical data analysis and just do a designed experiment for causation. In other cases, you could go directly to optimization because optimization cannot occur without causation, which cannot occur without correlation. You would have to be sure that the few factors you chose were root causes and it was just a matter of finding their optimal settings. This analysis is a specific DOE called *response surface*. Skipping correlation or causation has its risks, which should be considered when making the decision.

To establish causation, you have to control the factors. So, now you can understand why for suboptimal settings, you must have factors you are *willing* and *able* to control. Otherwise, you cannot show causation.

Correlation[3]

At every opportunity, you need the values of the settings with which you are trying to show correlation with percent good. Correlation is simply testing whether as one factor changed another did as well. So, you might check whether when you change the octane level of your gas, you get better or worse mileage. If you do, then the two are correlated. If you recorded the octane level and your mileage for several tanks of gas, you might be led to this conclusion. However, if you did not control for speed, weather conditions, traffic conditions, and so forth, the best you could conclude is that there is a correlation.

One also can use logic. You can group all the good opportunities and see what is common among them. Do the same for the bad opportunities, then try to discover what is different between the good commonalities and the bad commonalities.

Graphical analyses are wonderful and powerful tools as well for showing correlation. In fact, when it comes to showing correlation, it is far better to do the analysis graphically. Statistical results can be misleading in at least two ways:

- Statistics can indicate a strong correlation when, in fact, there is little correlation. The analysis was biased by a single point.
- Statistics can indicate no or weak correlation when, in fact, there is a strong relationship. A nonlinear correlation analysis is not revealed when a linear correlation analysis is done.

In each case, the graph will reveal this, the statistics won't. Use graphs. Then let the statistics quantify your interpretation.

Causation[4]

Causation is established by controlling factors and studying different combinations to see which ones affect a change in performance. The experiment is done by fixing some factors that might interfere in the study, not controlling others that you think are inconsequential, and then varying to specific choices the factors that you want to test for causation. These studies look at some or all possible combinations of the controlled factors and their choices. It can be expensive and time-consuming and typically involves statistical software to design the experiment and analyze the results.

You also can use simpler studies that rely more obviously on logic and/or simpler statistical analysis, e.g., Kepner–Tregoe and Shainin methods of problem solving. For example, you could select one good and one bad opportunity occurring at approximately the same time. Repeat this until you have half a dozen or so such pairs. Then, note the differences between the good and the bad in each pair followed by what is common in these differences across all pairs.

Optimization[5]

Once you know the factors that are causing defects, you want to make sure that the settings are optimal. If you find that as temperature increases, the number of defects increase and then at some point they start decreasing, you want to make sure you have the right temperature or range of temperatures that produces the fewest, if any, defects. The one temperature or range of temperatures is the specification or optimal setting.

The only way to do this efficiently and reliably is to test multiple choices for your factors, e.g., multiple temperatures. A study that does this while controlling for other factors is also a DOE-called response surface. It can be expensive and time-consuming and typically involves statistical software to design the experiment and analyze the results.

However, when you are looking at finding a local optimum and are considering only procedures that may be suboptimal or substandard, the statistical analysis is greatly simplified. The only analysis is to statistically test whether there are any significant differences among the percentages you already calculated for each procedure you are comparing.

You can use the chi-square test whether you have two or more procedures to compare. There are many free links that offer easy-to-use forms to do the calculations. Search by the test name and include "form" or "calculator" and "free" in your search.

If the test shows statistical significance, you have two cases:

1. One or a few are significantly better than the rest, so you have found a best practice(s) to standardize to.
2. One or a few are significantly worse than the rest, so you have found worst practice(s) to eliminate by standardizing to one of the others.

If you have more than one choice of procedure to standardize to, pick one based on cost, ease of use, ease of learning, process time, and/or other considerations. Okay, you can have both possibilities as well—treat it as case (1).

In the situation where suboptimality is due to the procedures used by people, you can at least identify which procedure is best. For example, five people do the same job, yet they do it differently. According to one measure, e.g., timeliness or completeness, the results vary significantly. Your task is to determine which procedure is locally optimal. Local optimality is simply which of these is best and not what possible procedure (including none of these) is best.

You will already have collected data for baseline performance on each procedure. Sometimes, just describing the differences among the procedures can inform you of better practices.

So, if you do not have descriptions of the procedures or they are not complete, now is the time to get the following information on each:

- Standard operating procedure
- Steps/activities
- Information
- People
- Tools
- Forms
- Databases
- Materials
- Consumables

If a procedure consists of separate tasks that are done in different ways, you may create a best procedure by determining which procedure is best for each task and create a "super" best procedure from these. This requires doing as many analyses as you have separate tasks.

Unpredictability

Review Chapter 4 (Unpredictability-Caused Defects). For unpredictability, data on percent good are still required, as with any other root cause. However, by definition, you are not finding root causes. Instead, you are finding factors that can predict percent good better than you are currently doing. Subject matter experts are your source for identifying potential predictors.

These predictors are also metrics, so describe them completely as you would for suboptimal settings:

- A label or name (often the labels are X1, X2, etc.)
- A definition

- The metric (list of possible choices)
- For quantitative factors, the resolution (smallest amount measurable)

The difference is that with these factors you are either unwilling or unable to control, or both.

Because you are not finding root causes but predictors, your analysis consists of developing mathematical models that show the relationship between the predictors and percent good.[6] There is no guarantee that you will find such a model that is useful. The usefulness of the model depends on two criteria:

- Can you predict within a predefined margin of error?
- Can you predict at a higher percent good than without the model?

You need a "yes" answer to both.

It can be time-consuming and involve statistical software to analyze multiple factors, especially when you consider complicated relationships.

Personal Reason

Review Chapter 5 (Personal Reason-Caused Defects). Personal reasons result in the person not complying. The noncompliance causes a defect. You need sufficient detail of the reason for you to determine why the noncompliance occurred. Then, you have two decisions to make for each personal reason provided:

1. Does the reason identify one of the other defect types (e.g., the reason "I won't pay an invoice is because it is incorrect")? If it does, then treat the problem for that type of root cause, e.g., due to a delay, an error, a suboptimal setting, or unpredictability.
2. If it is not due to a defect, then are you willing and able to address the reason? Those that you are not, inform the person and move on; those that you are, move to the next critical thinking question.

Remember, if the reason for noncompliance is because of a defect, then this type of problem converts to the type of defect stated. If you are either unwilling or unable to address the reason, then you cannot solve the problem.

Measurement Error

Remember that measurements are simply a way to distinguish between things. Sometimes you have qualitative choices and sometimes you have quantitative. It doesn't matter which; measuring selects one of the choices for the thing you are measuring.

When determining the percent good, you will need to make two measurements. The first is to identify whether the unit (product or service) is the type in which you are interested. This might seem so simple that it is not worth discussing. You might think: "How could anyone not know what it is that they want to look at or study?"

Let's consider some examples.

- Suppose you want to look at all current customers. Which of these is such a customer: One who has not purchased in one day, one week, one month, one year, five years?
- You want to determine how long it takes to send an invoice. Do you consider purchases that were canceled? Returned? Modified?
- The sales process requires senior executives to approve proposals. What constitutes an approved proposal?
- The goal is to have customers accept approved proposals. Is a proposal accepted if the customer says yes with a minor change? If the customer says yes, but later backs out? How long do you wait before deciding that the customer has not accepted the proposal?

The second measurement you need to make is to determine whether the product/service unit is good (or bad). Whether you are selecting from a limited number of choices or using an instrument to measure and then comparing to specifications, you could make a mistake. This mistake is also a measurement error.

The point is that you can make a mistake when selecting the product/service unit. Such a mistake is called a measurement error. These mistakes will occur when there are no operational definitions of the product/service, the characteristic, the measure, and the specifications. To reduce measurement error, you will need to define them. There are also analyses you can do to determine the extent of measurement error and how to reduce it. These topics are beyond the level of this book, but there are many resources on them. Search the Web for measurement error, gage r&r (GR&R), MSA (measurement system analysis), and also contact AIAG

(Automotive Industry Action Group) and Lean/Six Sigma groups and forums on the topic.

Summary

To answer the critical thinking question of how do you know the root cause, you need three pieces of information:

- Problem type:
  - Delay, error, suboptimal, unpredictability, personal reason
- Causal Relationship
- Data on causal relationship and process performance:
  - Delay: What is the process waiting for?
  - Error: Trace to sources
  - Suboptimal setting including substandard procedure: Critical factors/actions and their settings
  - Unpredictability: Predictive factors and their settings
  - Personal reason: No analysis—just ask

Table 7.1 summarizes the information on possible causes you need by problem type. Table 7.2 summarizes the analysis, if any, needed to determine root causes.

Because you have specified the information that identifies the root cause by type, the search for root causes is greatly simplified and, for some

Table 7.1 Information on Root Cause by Type of Problem

Root Cause	*Information*
Delay	Type of delay, duration, where/when it occurs.
Error	Type of error, frequency, where/when it occurs.
Suboptimal Setting	(Suboptimal machine equipment, etc.) For each factor that you are willing and able to control: label, definition, choices, resolution. (Suboptimal procedure) Description including steps, tools, forms, databases, materials, etc.
Unpredictability	For each predictor: label, definition, choices, resolution.
Personal Reason	The reason and, if it's not "due to a defect," whether you are willing and able to address it.

Table 7.2 Analysis of Information to Determine Root Cause by Type of Problem

Root Cause	*Analysis*
Delay	None unless not addressing all delays, then prioritize.
Error	None unless not addressing all errors, then prioritize.
Suboptimal Setting	(Machine equipment, etc.) Using historical data, identify which factors are correlated with percent good. Show if there is a cause–effect relationship between correlated factors and percent good. Determine the optimal settings (choices or specifications) for the factors proved to be causes. (Procedures) Statistically test whether there is a significant difference in percent good among the procedures.
Unpredictability	If possible develop a prediction model between factors and percent good with an acceptable margin of error.
Personal Reason	Determine if the reason is due to a defect. If so, treat it as such. If not, determine whether you are willing and able to address it. If so, do so. If not, inform appropriate person.

root cause types, almost eliminated. The basis for this is the cause–effect relationship that was established in the first section: Specific causes have specific effects. Learn what the effect looks like for each specific cause.

Endnotes

1. However, the use of the notation Y is often inconsistent. Sometimes the measure of the characteristic is labeled Y and the measure of the root cause is labeled X. Other times Y refers to the characteristic or the performance measure. For example, Y could refer to timeliness (characteristic), actual time to deliver in minutes (measure), or percent of on-time deliveries (performance measure). Similarly, X also is used to refer to either the cause or a measure of the cause. When someone uses Ys and Xs, ask them which use they mean, and be ready to see blank faces.
2. Measuring is a process and we want the measurements to be reliable so we can trust the results. We don't want to think we have a problem when we don't or vice versa. Reliability of your percent good calculation depends on your ability to (1) distinguish good from bad and (2) count the number good. Statistics can help you do measurement system analysis to evaluate the reliability of your measurements. This analysis is used to evaluate the accuracy, precision, stability, linearity, and detection limit of your

measurement process. Depending on how poorly your business process is performing, you may be able to do an analysis that is crude but adequate. For example if you want to reduce cycle time from months to weeks or days, just ballparking how long delays are will be enough.

3. Statistical hypothesis tests are used to establish correlation with historical data. There are different tests you can use that depend on whether your opportunity metric Y is quantitative continuous or not (discrete) and whether your root cause metric X is quantitative or not. Some tests are summarized as:

   Y, X continuous: Modeling and statistical
   Y discrete, X continuous: % good
   Y continuous, X discrete: means, variation, % good
   Y, X discrete: % good

   However, some software allows you to do these analyses without necessarily understanding all the tests. The software based on the type of Ys and Xs you have can find the correlations and estimate optimal settings, if the factors are also causally related.

4. There are different kinds of designed experiments from full factorials, which look at every possible combination, and fractional factorials, which look at a specific portion of all possible combinations.

   For example, you could test your theory about octane and gas mileage by driving the same distance on the same road under the same weather conditions. Perhaps you also wanted to test the speed at which you drove. So, you do the drive for each of the six combinations (if done correctly, you would also randomize the order in which tested the combinations):

   Octane = 92, speed = 65
   Octane = 92, speed = 55
   Octane = 89, speed = 65
   Octane = 89, speed = 55
   Octane = 87, speed = 65
   Octane = 87, speed = 55.

   While the results of these design of experiments (DOEs) are analyzed statistically, the statistical analysis is really quantifying logic. You can use logic to also determine root causes. The logical approach is to compare and contrast good opportunities and bad opportunities. This is the kind of logic that is used in puzzles, e.g., sudoku, and what you might see in detective stories, e.g., TV's *Crime Scene Investigation*.

5. In the mileage example, let's say that you discovered that it wasn't octane but speed that made the biggest impact. So, next you want to know what speed will maximize your gas mileage. Now you can pick one octane and the same road and conditions, and choose three or more speeds at which to drive. This will tell you which speed is optimal. The more complicated the relationship between the root cause and performance, the more values you may have to test. And, when there is a combination of factors that work together (called an interaction) to get the highest percent good, the experiment and analysis becomes even more complicated.

6. Start with linear relationships of single factors. Then try linear relationships of multiple factors followed by nonlinear relationships with single factors and, finally, nonlinear relationships with multiple factors. Mathematical skills and appropriate software are typically required for this type of analysis because the equations developed for the prediction models depend on whether the Ys and Xs are quantitative or not. The trick is to find the simplest model that will serve the purpose of predicting within the margin of error at an acceptable percent good. There are tools for transforming data to make the models simpler, but with today's computer capabilities, it almost never makes sense to do so.

Chapter 8

How Does the Proposed Solution Work?

The proposed solution works because you know that you can sustain the closed gap with the proposed process change.

The information you need includes:

- A proposed process change (solution)
- Confirmation that the change closes the process performance gap
- Controls to sustain the improved process performance

You already learned in Section I how to identify process changes that are solutions by problem type. Some of the cases also included examples of how others have included changes aimed at sustaining the improved process performance.

To find solutions, for each requirement, review the How to Address section in the five chapters in Section 1. Because you do not want the defect rate to revert to the baseline level or deteriorate from the new level of performance, you include in solutions methods of sustaining the new performance level. These methods of sustaining are based on the concept of control and continuous improvement.

Process Change to Close the Gap

Improving the process requires that you be willing and able to address the root causes. If so, then you have two tasks. The first task is to determine what to change in the process and the second task is to increase the likelihood the change is permanent and being used. The first task depends on the root cause, as shown next. The second task requires controls. Controls range from mistake-proofing to monitors. Use mistake-proofing whenever possible.

Process Changes with Mistake-Proofing to Address Root Cause

Delays—(not due to defects or slow processes) are addressed by creating continuous flow, reducing distances and motion, transporting faster or not transporting, leveling demand to meet supply, eliminating steps and tasks, simplifying procedures. Making tasks and sequences simpler and easier is low-level mistake-proofing.

Errors—are addressed by mistake-proofing. Recall this sequence in Chapter 2.

Errors cause Defects cause Consequences

Given this sequence, there are four places you can intervene, in the order of from worst to best:

1. After the consequences
2. After the defect, but before the consequence
3. After the error, but before the defect
4. Before the error

Case (1) requires mitigating the results. Cases (3) and (2) are attempts to block with 100% inspection between (2) and (1), though it will not catch all defects. Case (4) is mistake-proofing. See *Next Steps* in Chapter 2 for more information on mistake-proofing.

If you cannot completely mistake-proof, consider other tools in the Lean arsenal. A simple but effective approach is the 5S (sort, set in order, shine, standardize, and sustain).

Sort so that only what is necessary is available. Set in order by having a place for each thing and each thing in its place. Shine by keeping work area clean. Standardize work. One way to standardize is use checklists. Sustain these practices (which is why *sustain* is included in the third critical thinking question and why there is a fourth critical thinking question). Adopting these practices reduces the chance of using the wrong tool or form.

Suboptimal setting—While this root cause is the most complicated to determine, it is easy to know what to change. Because you have only selected settings you are willing and able to control and the analyses told you the specifications for those settings, you just have to put the critical settings within their specifications. Mistake-proof to prevent the wrong settings by locking the settings or using labels, colors, and other visual aids.

When the suboptimality is due to substandard procedures and you are only looking to locally optimize, then the analysis will be simpler. Substandard procedures are addressed by finding best procedures and standardizing to one of them (using cost, ability to train others, time, and other factors to resolve ties in performance). To reduce the likelihood of substandardization creeping into the new standard operating procedure (SOP), consider training and testing all new employees with a standard training method and removing ways (tools, materials, forms) of doing tasks differently or making it burdensome to do so.

Unpredictability—The analysis told you what the mathematical relationship is between the leading indicators (predictors) and percent good. Now you have to develop a standardized operating procedure for using the model. That procedure will include when it is used and what actions to take for each possible prediction or range of predictions. This may require training on the model and the procedure, stating conditions or restrictions on when the model is to be used. Use check sheets, guides, visual aids, software screens, etc.

Personal Reason—The only personal reasons that get to the Improve step are those that (1) you have identified as not related to a defect and (2) you are willing and able to address. Your task is to determine how you will address the reason, which should be relatively easy given you have already judged that you can. If you only need to address the reason once, then you are done. If you need to address the reason periodically, then find ways of mistake-proofing. For example, if a customer does not renew because he/she forgets due to the length of time between renewals, then create an automatic reminder. Develop an individualized solution forward with the person.

Process Changes with Controls to Sustain Improved Performance

The purpose of a control is to (1) tell you in a timely manner when performance has changed and (2) indicate what action to take. A good control will sustain the new performance level.

The best way to manage the effectiveness of your improvement solution is to take these two steps:

1. Mistake-proof as much as possible.
2. Add controls if the highest level of mistake-proofing cannot be done.

Above, you saw how to include mistake-proofing in your solution to address the root cause. Controls address your solution.

From Section I, you learned that there are levels of mistake-proofing. If you can prevent the defect from occurring, that is wonderful. However, for a variety of reasons, including cost, you might not be able to or don't wish to install the highest level of mistake-proofing.

Without mistake-proofing, processes will change their performance for many reasons including degradation, changes in factors that may not be in your control, and changes in factors that did not occur to you as being influential or you thought they were not influential. A control is a method for tackling such changes.

A control has three parts:

- Measure
- Signal
- Response

Table 8.1 Controls by Type of Problem

Root Cause	*Analysis*
Delay	Monitor total time (effect), time for selected tasks (effect), and whether changes are followed (cause)
Error	Audit mistake-proofing (cause) to ensure still prevents errors (effect)
Suboptimal Setting	Monitor performance (effect) and settings (causes); for suboptimal procedures, test for proficiency (cause), monitor performance (effect), monitor SOP followed (cause)
Unpredictability	Audit to ensure the procedure is used (cause); make periodic analyses to confirm model is still valid (effect)
Personal Reason	Periodically ask the person if reason has been/still is addressed (cause); audit that the change is still in place (effect)

The measure is either of the product/service characteristic or of the causes. It is better to put a control on the cause rather than the effect. If you only track the effect, then you will only know after the fact that there has been a change.

The signal is an indication that something changed. The response is an action addressing the change. Table 8.1 summarizes controls you can use for each type of problem.

A good control inculcates a culture of not passing defects.

Summary

There is one deliverable in answering the third critical thinking question: a sustainable solution. However, that solution must address two issues:

- How to close the gap between baseline and desired performances
- How to provide signals of process performance worsening

The information you need is:

- A proposed process change (solution)
- Confirmation that the change closes the process performance gap
- Controls that can sustain the improved process performance

The fourth critical thinking question is covered in Chapter 9 and illustrates how to use the controls you have installed in the solution to ensure sustainment of the new process performance. In addition, controls provide opportunities for further improvement.

Chapter 9

How to Know When to Improve Again

You know you can improve again because there is an unplanned change in process performance for which you can find the root cause and benefit.

You can always set a higher desired process performance level, which is worth closing the gap. This is deliberate and scheduled improvement. But why wait? Opportunities often occur for improving a process that are not scheduled or planned.

The key is knowing when these opportunities occur. The information you need includes:

- A signal that indicates process performance has improved
- A method for root cause analysis

You already have this.

The controls you put in place have the signal:

- Measure
- Signal
- Response

Recall that the better measure for sustaining process performance is of the cause and not of the effect. However, for improving, the measure of the effect is needed to know whether process performance improved. A change in process performance can be either a fortuitous improvement or an unfortunate worsening of performance.

You already know two ways to find and address root causes: the three-step procedure for single output units and critical thinking for multiple output units.

To continuously improve, you must do two things:

- Sustain the current process performance
- Take advantage of opportunities of unplanned changes in process performance

Controls are the key to doing both.

A control uses criteria to determine whether a change occurred. The criteria help distinguish between changes and no changes. The signal tells you that something has changed. It doesn't matter whether the change is for the better or for the worse, you want to know when it occurs. The signal is a trigger for one of three responses:

- If performance improves, take preventive action to keep it at that better level.
- If your performance is steady, continue what you are doing.
- If the performance worsens, take preventive action to return to previous better level.

Preventive action consists of using critical thinking and the three-step procedure.

A control indicates that one of three possible changes occurred:

- A defect occurs
- Process performance[1] changed
- Process stability changed

Defect Occurs

When a defect occurs, this applies to a single output unit. Section I showed you how to deal with this situation.

There are two responses you can take. The first is corrective action to either rework or replace the defective unit. Follow the Lean principle and don't pass on the defect. Second, feed this information about the defective unit into a preventive action procedure, namely, the three-step procedure of Section I using the second and third critical thinking questions.

The preventive action can accomplish both objectives of continuous improvement. By addressing the root cause of this defect, you may only return process performance to its desired level. This action will then sustain the current performance.

But, if you reduced the defect rate, you improved process performance. Then the action will improve the performance through an unplanned opportunity. You won't confirm whether you have only sustained or improved process performance until you calculate process performance after the preventive action.

Change in Process Performance

A change in process performance applies to multiple-unit performance. Recall that process performance is the percent good, thus many units are involved.

When the signal indicates process performance worsened, you need to respond to bring it back to the previous performance level. Check whether defects have occurred. If they have, take corrective action.

Take preventive action to get the process back at least to the same level as before this change. You can start with step two of the three-step procedure and answer the third critical thinking question. You can start here because of what you already know. You know the customer requirements for these output units. You know the type of defect that can result when the requirements are not met. You know the baseline performance, the level of the changed performance, and whether you have a gap. You only need to assess whether the gap is worth closing.

If you return the process to its performance level before the change, then your preventive action sustained the current performance.

Sometimes process performance fortuitously improves. A change for the better gives you the opportunity for unplanned improvement. Your response is to get the last piece of information to answer the critical thinking question: "How do I know I have a process performance problem?" The gap now is between the better process performance that is indicated by the signal and the current process performance. Take these steps to benefit from this opportunity to improve:

- Determine whether the gap is worth closing to answer: "How do I know I have a process performance problem?"

- If the gap is worth closing, answer: "How do I know the root cause?"
- Then answer: "How do I know my solution works?"

In other words, you can shorten the problem-solving procedure because you already know most of the information to answer the first critical thinking question. This preventive action improves process performance.

Change in Stability

Instability also applies to multiple units. An unstable process requires much attention, so it is more efficient to have a stable process. A process that is stable is said to have common cause variation. Signals for instability are called special cause variation. Like signals for process performance, they can be good or bad. A good signal or good special cause variation indicates that the process has improved. A bad signal or bad special cause variation indicates that the process has worsened.

If the instability created defects, take corrective action on the defective output units. For instability that worsens process performance (a bad signal), take preventive action to return to the current performance. This is a sustaining action.

If the instability improved process performance (a good signal), then take preventive action to change the process performance to this new level. The purpose of this problem-solving activity is to make the good special cause variation common cause variation. This will raise the level of performance to the level achieved by the special cause.

In both situations, you can shorten problem solving. You have all the information to answer the first critical thinking question ("How do I know I have a process performance problem?") except whether the gap is worth closing. Start here and continue answering the second and third critical thinking questions.

Summary

The purpose of the fourth and last critical thinking question is twofold:

- Take advantage of unplanned changes to sustain and improve on current process performance. All three types of changes (defect, process performance, stability) create opportunities to both sustain and improve.

An additional benefit of these opportunities is that you can shorten the time and effort to problem-solve because of the information you already have. You have fewer critical thinking questions to answer. The only information you need is

- Whether the gap is worth closing
- How to close the gap sustainably

Endnotes

1. In technical terms, process performance is also called capability. Often, people ask whether a process is capable. This format of the question suggests that the answer is yes or no. However, capability (and process performance) is quantitative. There are degrees of performance. As a percent good, it ranges from 0% to 100%. The goal statement states the minimum desired performance or capability. The better question is "Is the capability or process performance at the desired level?" Or, if you do not know the desired level, ask "What is the capability (process performance or percent good)?"

Chapter 10

Designing Processes

Principle

The third application of Lean's fifth principle to seek perfection is to create processes with few or no defects rather than reducing the defects after installing the processes. You will not be able to eliminate all defects. However, you can further reduce the probability of defects occurring by applying what you have learned to improve processes to designing processes.

In this chapter, you will learn how to create processes based on seeking perfection, thereby heeding principle 5. The level in this book is basic. While you learned about some more complex statistics when discussing suboptimality- and unpredictability-caused defects, analyses requiring other advanced mathematics and science (e.g., physics, chemistry, engineering, psychology) are beyond this book. Please check other sources for this material, e.g., do a Web search on Design for Six Sigma, statistical tolerancing, ideal value-stream map.

Methodology

You want a methodology for re-creating processes that start with a high percent good and are insensitive to changes in the environment or other uncontrollable factors. The less effect the uncontrollable factors have on performance, the more robust the process and products are.

The critical thinking question for designing processes is:

How do I know my process will be robust and high-performing?
The answer is: When you know the process performance has passed performance and robustness tests.

The information you need consists of the following for all five problem types:

- Desired process performance levels
- Mistake-proofing/controls in the process design
- Evidence that process performance meets desired performance levels under various conditions
- Evidence that mistake-proofing/control works

Fortunately, for designing processes, you can continue using the critical thinking questions for multiple-unit process improvement. There will be three changes in the details, but not in the purpose of answering the questions. The purpose and information needed for each question stay the same.

How Do I Know I Have a Process Performance Problem?

The purpose of this question is the same—to determine whether you have a gap that is worth closing. The information also is the same:

- The gap between current and desired process performance:
  - Current process performance:
    - Requirement
  - Desired process performance
- The value of closing the gap:
  - Consequence or benefit

For new products and services, you will need to learn from potential customers what all the output requirements are. Typically, this comes from market analyses. Gather and record the data (perhaps using a template that has the four customer requirement components) to capture and organize the requirement information.

From the customer, you will need to know how critical these requirements are relative to each other. Add another column in the template table (e.g., Chapter 6, Table 6.1) that show the relative ranking of these requirements according to the customer.

Typical rankings use a 1 to 10 scale. Another approach to use instead of or in addition to is to determine what the consequences are to the customer and business as a result of failing to meet a customer requirement on a single unit and chronically on multiple units.

When designing new products and services, the process that produces these products and services does not exist. You do not have data for a baseline. However, you can state that the current performance is 0% good (or 0 for whatever measure you are using).

You must state the desired performance levels for all the requirements the output unit is supposed to satisfy. Because the desired performance will be greater than zero, then you know you have a gap. It is not sufficient to say that a process does not exist. You need to state the performance desired for each requirement:

Example of Baseline Performance Statement: Today (*date*) there is no process for delivering service X to market Y with performance levels on quality of xx% good, cost of $xx, and speed of xx/period.

Example of Desired Performance Statement: By (*date*), have a process for delivering service X to market Y with performance levels on quality of xx% good, cost of $xx, and speed of xx/period.

As before, the company decides on the process performance levels.

Since the company has decided to design and install this process, it must have already assessed that closing the gap was worth it. But, to ensure clarity and understanding, make sure that the business impact statement references the desired performance levels and not arbitrary gains.

How Do I Know the Root Cause?

The purpose of this question is the same—to determine what causes the gap. The information also is the same:

- Problem type (for efficient searches of causal relationships)
- Causal relationships by problem type
- Evidence confirming the causal relationship to process performance

You already know how to and have experience determining the problem types that are possible from the requirement. You should do that first to answer the second critical thinking question.

As with the current process performance, you also do not have data on potential root causes. Because neither the process nor the output exist, these data will be generated through research and development using simulations, studies, scientific laws, prototypes. These data also will be on potential defects, because you do not have output units. You are not analyzing to find root causes of actual defects, but you are analyzing to find root causes of potential defects.

One benefit of knowing about the five types of process problems is that you can organize your search for these potential root causes using this information. Because you know how to address existing defects due to these five types of root causes, you can design the process to prevent or drastically reduce the probability of their occurrence. For each type of root cause process problems (delay, error, personal reason, unpredictability, and suboptimal setting), determine where they potentially can occur in the new process. Since you now know what the symptoms of these root causes are, you will find it easier to identify them.

The following questions will help you search systematically:

- Timeliness: How will you create continuous flow?
- Error: How can you mistake-proof?
- Suboptimality: What are the optimal settings for machines including software? If multiple people will be doing same task, what's the best standard operating procedure (SOP) and how will you ensure everyone uses it? How will new people know the SOP and be proficient with it?
- Unpredictability: Where do actions and decisions depend on forecasts or predictions?
- Personal reason: Where in the process is someone likely to do something he/she is not supposed to do or not do something they are supposed to do?

Create a table (Table 10.1) with the ordered activities in one column and the five types of causes in other columns. Include an appropriate question for each cause type. Note whether the defect associated with that cause can occur with the activity. Then develop answers to the specific question.

Table 10.1 Designing Processes Without the Five Process Problems

PROCESS STEP OR ACTIVITY	*DELAY: How will I create continuous flow?*	*ERROR: What can I mistake-proof?*	*SUBOPTIMALITY: What are the optimal settings? What steps must be standardized?*	*UNPREDICTABILITY: What forecast do I need before I proceed?*	*PERSONAL REASON: Is what I consider a defect also a defect for others?*

Table 10.2 How to Identify Information Needed to Prevent the Five Process Problems

	INFORMATION	
PROCESS STEP OR ACTIVITY	*What tells me to start activity?*	*What information do I need to do activity?*

Critical thinking questions aim at identifying the information necessary to solve process problems. Similarly, you can use the basis for critical thinking to design better processes by identifying the information necessary (1) for knowing when to do the activity and (2) to do the activity. Add two more columns under an overall heading of Information, one for when to do the activity and one for what information is needed to do the activity (Table 10.2).

How Do I Know I Have a Sustainable Solution?

The purpose of this question is the same—to determine sustaining ways to address the root causes. A fourth piece of information is added to the information for improving processes:

- A proposed process change (solution)
- Confirmation that the change closes the process performance gap (for all requirements)
- Evidence that the improved process performance can be sustained (for all requirements)
- Evidence that the process change is robust (for all requirements)

When finding solutions for process performance improvement, you looked for solutions to actual root causes of actual defects. When designing processes, you will be determining actual solutions to the potential root causes of potential defects. In both cases, your aim is to sustainably address the root causes.

The degree to which you do this depends on two issues. One issue is how effective your controls are. Whether improving existing processes or designing new ones, your solution will need controls. The development and implementation of controls is exactly the same whether it is for an improvement solution to an existing process or a new process.

The second issue is how robust your design is. Robustness means that process performance will be relatively consistent even with changes in uncontrollable factors. Such factors might include environment, personnel, raw materials, etc. The key task is exemplified by this question:

> How will you make the process robust to uncontrollable factors and what factors might need to have their effects predicted?

Finding an answer to the critical thinking question should be easier for you now. One benefit of knowing the five types of process problems is that you can organize your search for these potential root causes using this information. Below is a procedure to help you organize your search for solutions:

1. Use the requirement template that has for each requirement its defect type.
2. Determine what actions in the process will be value-added (VA).
3. Determine which VA actions can have errors; develop mistake-proofs for them.
4. Determine which VA actions need optimization; use design of experiment (DOE) on critical factors.
5. Determine which VA actions depend on predictability; use correlation analysis to identify predictors and develop a model; develop a procedure for using model and continual confirmation of its predictability.
6. Determine which VA actions could have defects for personal reasons; develop a procedure for asking and determining whether the reason is due to another type of defect or is a real personal reason; develop criteria for determining which real personal reasons you are willing and able to address.
7. If you cannot design a process with only VA actions, determine the amount of delay for the nonvalue-added (NVA) actions; develop a procedure for assessing whether NVA actions are increasing delay time.

Sustainable solutions maintain current levels of performance. For existing processes, this is done after installing the improvement solution. For designing processes, this is done after installing the new process.

The deliverable for Sustain is the continued maintenance of performance at the desired levels as opportunities arise. This also stays the same whether applied to existing or new processes.

You also will generate data to confirm your solutions robustly, and sustainably achieve the desired performance levels. Manufacturing processes do go through prototype and pilot testing before being installed full scale. Statistical analyses are typically required for these tests. The purpose is to determine the optimal settings of the machinery and optimal characteristics of raw materials. In regulated environments, there are procedures and sampling plans required for validation of the processes.

Service processes, unfortunately, do not go through this rigor. Too often, service processes may not even be piloted before being installed. For example, the IT Department may require a form to be used for requests for its services and the form is not tested for its effect on delays and errors in requests. Service processes can benefit as well from applying statistical analyses to its prototype processes before being installed.

How Do I Know When I Can Improve Again?

The purpose of this question is to take advantage of unplanned opportunities to exceed current levels of performance. The information you need stays the same:

- A signal that indicates process performance has improved.
- A method for root cause analysis.

For existing processes, this is done after installing the improvement solution. For designing processes, this is done after installing the new process. Once the new process is installed, it is no different than an existing process. You do the same thing to answer this question in both situations: Respond to the signals indicating an improvement opportunity exists.

TEACHING PROBLEM SOLVING

Principle

In the introduction to Section II, you learned that the principle to seek perfection had several applications. One was to continuously improve. This application is embedded in the procedure for improving processes as the fourth critical thinking question. The second application was in designing new processes that are robust, sustainable, and with very low probabilities of producing defects. The third application is teaching others how to solve these five types of process problems effectively and efficiently.

Teaching also provides an opportunity to create a culture of continuous improvement if done explicitly for that purpose. A teaching environment will support and enhance the purpose of creating such a culture under three conditions

- Teaching should be based on learning theories that support the cultural behaviors desired.
- Teaching should mimic the culture that is to be created.
- Teaching should embed the thought processes of the desired culture in the procedures.

Learning Theories

Why do some people do much better than others? There are many reasons. Certainly some naturally have more skills or aptitude than others. Some practice more than others. While no one cannot guarantee that everyone will be a star or superior problem solver, teaching in certain ways can help students be better, sooner.

The book *Blink: The Power of Thinking without Thinking* (Little, Brown Publishers, 2005) reports how experts have natural instincts. In the blink of an eye, they assess a situation and make a decision. (The author, Malcolm Gladwell, doesn't provide data indicating how often these decisions are correct, better, or even good. However, other research does provide support for the premise of the book.) This is epitomized in the two-by-two table or hierarchy of competence and consciousness:

- Level 1: Unconsciously incompetent
- Level 2: Consciously incompetent
- Level 3: Consciously competent
- Level 4: Unconsciously competent Blinkers are at Level 4: unconsciously competent.

Studies also show that many hours of practice alone does not an expert make. Yes, many hours of practice are necessary, but not just *any* practice. When the practice is specific to the areas that need improving, the practice becomes significantly more helpful. Even more helpful is when the practice provides immediate feedback indicating whether it is good or bad, correct or incorrect. This type of practice—focusing on weaknesses with immediate feedback—is called deliberate practice.[1] Hours and hours of such practice are necessary to reach the level of expert.

In sports, it is easy to see examples of deliberate practice. Tiger Woods dominated the golf world early in his career. But when he wanted to go beyond that, he sacrificed (did not win a major tournament in 2003 and 2004) nearly two years while focusing on what he believed were his weaknesses. His "second" career made him even better. He is now on his "third" career, after an even longer break of not only not winning any major tournament, but of not even winning any tournaments. As of mid-July 2012, he ranked No. 1 in the FedEx Cup.

LeBron James, after his Miami Heat suffered a humiliating defeat by the Dallas Mavericks in the 2011 NBA Finals, worked on his weaknesses as a player and a person. The result was come-from-behind victories against Indiana and Boston and a resounding win in the 2012 NBA Finals where he was named MVP.

Deliberate practice leads to the eventual blink phenomenon or the instinctual response. As more recent studies show, superior pattern recognition ability with many repetitions on the right things lead to deeper understanding, instinctual habit of recognizing patterns, and superior performance. The "key words" listings in Chapters 2 through 6 provide clues on how to distinguish among the patterns.

This book alone will not make you an expert. Perhaps the deliberate practice with immediate feedback won't either. However, the combination can make you good or even very good.

Purpose of Teaching

What: To acquire skills in solving process problems more effectively and efficiently.
How: (Deliberate) Practice, practice, practice—on pattern recognition with immediate feedback.
Environment: Make mistakes. Learn from them as soon after as possible. Practice again.

It's that simple.

Course Structure for Optimal Learning

Empirical studies show that practice and learning from mistakes provide the best learning opportunities for skill acquisition. However, there also are four other conditions that increase the permanency of learning.

Repetitive unpressured testing helps you retain what you have learned. One reason is that it gives you practice at retrieving what you have stored and identifies what you have not stored. You learn from this "mistake" what you don't know. In other words, you move from unconscious incompetence to conscious incompetence. When coupled with immediate feedback of what is correct and why rather than merely grading, you can accelerate learning. Lessons based on practice–feedback loops are a key.

A second condition is to learn things in different contexts and in variations. Apparently this helps because you learn to recognize and distinguish the specific issue from others. When it comes to applying what you learned, you are better equipped to make necessary distinctions because you have practiced separating the one specific item from its background. You move from conscious incompetence to conscious competence.

Thirdly, build on previous learning. It is much harder to retain and acquire skills when there is a mix of topics. Don't mix other topics that are lecture or examples only. They do not provide practice or feedback and, as a result, they do not teach skills.[2]

Finally, provide breaks between lessons on the same topic. Give students time for their minds to assimilate what they have learned. Starting another lesson with a type of test or review or practice combines these different conditions. Practice, practice, practice will move you from consciously competent to unconsciously competent.

Empirically Based Teaching Philosophy

Here are some guidelines for creating courses:

- Start with learning objectives that state what specific skills the student will acquire *in class*: At the end of this class/lesson/activity, the student should be able to *describe the activity* then *describe the environment and conditions.*
- Allocate at least 75% of the time on practice with feedback.
- Ensure participants go through at least two cycles of learn–practice–feedback–practice.
- Feedback should follow this format:
  - If the practice is correct, acknowledge and provide another practice that is more difficult.
  - If the practice is not completely correct, review the mistake(s) and provide practice on the same issue at the same level of difficulty.
- Practice on problems they encounter and processes they use. Have them bring their real problems. Be wary of simulations. Often they have artificial problems that frequently are solved without the tools taught.
- Start simply and then build on what they already know.
- Practice on a few simple skills and add more skills and complexity as they master the previous skills.

The chapters in this section provide guidance for creating teaching activities that comply with the six conditions that support creating a culture of continuous improvement through teaching:

- practice
- learning from mistakes
- repetitive unpressured testing

- learn in varied contexts
- build on previous learnings
- provide breaks between lessons

The 14 lessons in this section are separated into two groups. Chapter 11 has seven lessons covering the material of Section I, based on teaching the three-step procedure. It also includes a lesson on requirements. Chapter 12 has seven lessons covering the material of Section II, based on teaching critical thinking for chronic problems and designing processes. Chapter 13 provides guidance for mentoring and developing mentors.

Course Structure

Courses consist of lessons, which have a purpose and provide practice on a single topic. In Chapter 11, seven lessons are described from which courses combining the lessons are suggested. The content of each lesson includes:

- Topic
- Learning objectives
- Prerequisite
- Class length
- Materials
- Preparation
- Activities (learning structure)

The learning structure consists of the methods used to transfer knowledge and skills. The activities used determine the learning structure. Activities based on learning theories should provide the most benefit and greatest transfer and retaining of knowledge and skills.

There are five types of activities or exercises: (1) pattern recognition with immediate feedback, (2) practice on producing deliverables, (3) writing (e.g., descriptions, documenting results, explaining why), (4) discussions (e.g., what went well, what was unclear or difficult, explaining, critiquing the work of others), and (5) teaching. Every lesson includes the first four and several include the fifth. These are explained in the activities.

Curricula

Courses are designed by combining lessons. Sets of courses make a curriculum.

Lessons 1 and 2 teach the fundamentals of this book: the five types of process problems and their relationship to requirements. Lessons 3 to 7 teach the application of the three-step procedure to each of the five types of process problems based on single units. Lesson 8 teaches critical thinking. Lessons 9 to 13 teach the application of critical thinking to chronic problems on multiple units. Lesson 14 teaches the application of critical thinking to designing processes.

To adhere to the learning principle of building on existing knowledge, there are three levels of courses you can design. This curriculum is based on skill level. The basic level would consist of Lessons 1 to 4 and 7, thus teaching the fundamentals and application of the three-step procedure to the most common problems: delays and errors.

The intermediate level course would consist of Lessons 8 to 10 and 7. This would teach the full, four-step procedure of critical thinking questions and how to apply it to the most common process problems. Lesson 7 can be added as a review of the three-step procedure and application to personal reasons that are caused by other types of defects.

The advanced level course would include the remaining lessons, Lessons 5 to 6, 11 to 13, and 14. The audience for these courses may be restricted as these defects are common in certain applications, but not in most.

Rather than creating levels of courses, a curriculum could focus on the topics. For example, three courses could be as follows (Curriculum II, Table III.1):

- Delays and Errors: Lessons 1–4, and 8–10
- Suboptimality and Unpredictability: Lessons 5–6 and 11–12
- Personal Reason and Design: Lessons 7, 13, and 14

Table III.1 summarizes two other possible curricula where the common letters denote groupings of courses for different audiences or levels. For example, Curriculum I has three courses: one on single unit process improvement, one on multiple-unit process improvement, and the third on designing processes. A fourth curriculum might separate Lesson 14 on process design from the others creating a fourth course. Other variations are possible.

Table III.1 Lessons for Creating Curricula

Lesson	*Topic*	*Curricula*			*Chapter*
		I	*II*	*III*	
1	Five types of process problems	A	A	A	11
2	Requirements	A	A	A	11
3	Three-Step Procedure for Delays	A	A	A	11
4	Three-Step Procedure for Errors	A	A	A	11
5	Three-Step Procedure for Suboptimality	A	C	B	11
6	Three-Step Procedure for Unpredictability	A	C	B	11
7	Three-Step Procedure for Personal Reasons	A	A, B	C	11
8	Critical Thinking	B	B	A	12
9	Critical Thinking for Delays	B	B	A	12
10	Critical Thinking for Errors	B	B	A	12
11	Critical Thinking for Suboptimality	B	C	B	12
12	Critical Thinking for Unpredictability	B	C	B	12
13	Critical Thinking for Personal Reasons	B	C	C	12
14	Process Design	C	C	C	12

Suboptimality and Unpredictability Courses

These two courses should have a limited audience. You can ensure increasing the benefit of the course by having prerequisites. Prerequisites would include that the participant's job requires the regular use of design of experiment (DOE) for the suboptimality course and forecasting for the unpredictability course. Another requirement might be a certain level of mathematics and software capability.

You will have to select software to do the analyses for these courses. Then you have to decide whether you teach the software in class, provide online training, or specify which software tutorials are prerequisites for these courses.

Participants should bring descriptions of the process they want to optimize and forecast. However, using their own processes might not allow even one practice feedback loop in class. If they need to reset machines, for example, you will not be able to collect data. You could restrict the class to people operating the machines and hold it onsite.

Depending how far in advance the forecast is, you won't be able to complete a practice–feedback loop in class, either.

Unlike error and delay, creating suboptimality and unpredictability defects in the registration process is more difficult. Instead, these two courses are more suited to using simulations.

Make sure, however, that students have at least two practice–feedback loops in class. This means that, for suboptimality, the factors must be controllable and they must be able to run the process. For unpredictability, reality must occur before the class ends. In other words, the forecasts must occur before the class ends.

After at least two rounds of practice–feedback, go to their suboptimal or unpredictable processes to apply what they have learned. If feasible, schedule another half day for reviewing the results of their solutions.

Each course can be done in one day when it is restricted to just learning, practicing, and getting feedback on basic DOE and correlation analyses.

A better approach would be to bring the class (lessons) to the people as they work on their processes while at their job.

Personal Reason Course

You may find this course the hardest to promote. Some people may object (one person quoting W. Edwards Deming told me that people are never the problem) by claiming this is finding fault with people. It isn't.

Review the section Identifying Personal Reason Caused Defects in Chapter 5. Often the personal reason leads to the other kinds of defect causes. Understand that is not a defect from the person's point of view, but is from the process's view, e.g., employee or customer leaving the company.

This course can use the registration and attendance processes as the first example on which to work. If people do not show for the course after registering, it is viewed by the process as a defect, but not necessarily by the absent person. Stress this point.

Then ask whether it is an issue of error, delay, suboptimality, or unpredictability. Discuss how it could be these others by listing examples of situations that led to this defect.

Then ask how you can know. This should lead to asking the person.

When the reason does not lead to one of the other causes of defects, then you must determine whether you are willing and able to address the person's reasons.

Loops of practice–feedback can be done in pairs with one person providing a case where they had a personal reason and the other person determining the root cause. Switch roles and repeat.

Summary

There are several keys to successful transfer of knowledge including:

- Build new knowledge and skills on existing ones.
- Start with basics, then expand after knowledge and skills are acquired.
- Base learning on pattern recognition.
- Provide multiple loops of knowledge/skill–practice–immediate feedback.
- Provide multiple examples and practice on complete methods and procedures.

Endnotes

1. New research shows that outstanding performance is the product of years of deliberate practice and coaching, not of any innate talent or skill. (Ericsson, K. A., Prietula, M. J., and Cokely, E. T. 2007. The making of an expert. *Harvard Business Review*, July).
   Ericsson, K. A., Krampe, R. T., and Tesch-Römer, C. 1993. The role of deliberate practice in the acquisition of expert performance. *Psychological Review*, July 100 (3): 363–406.
   http://www.coachingmanagement.nl/The%20Making%20of%20an%20Expert.pdf
2. Every course I have seen disregards this condition by embedding in learning how to solve process problems' other topics, e.g., of project management, strategy deployment, and change management. The result is very little practice *in class* on solving problems.

Chapter 11

Three-Step Procedure Lessons

Lessons 1, 3–7 of this chapter follow the material in Section I. Lesson 2 covers requirements, which was presented in Section II. Whether you do Lesson 2 in the order described below or after Lesson 7, it needs to be a prerequisite to the Lessons 8–13. Use this book as a reference for class participants, including developing materials from it, e.g., chapter introductions, examples, tables, figures, etc.

Lesson 1: Five Process Problems

Topic: Five types of process problems

Learning Objectives: After the lesson, participants will be able to state, identify, and generate examples of the five types of process problems by their root causes and their effects.

Prerequisite: None.

Class Length: Two (2) hours.

Materials:

- An example of each type of process problem with a word description
- Electronic defect flash cards with answers (20–25)
- Electronic cause flash cards with answers (20–25)
- Problem Type Table

Preparation

- Prepare the **defect** flash cards by writing on each card a one-sentence description of a defect using the key words from Section I chapters. Create four to five per process problem type. Create a slide show with one defect per slide randomly ordered and after each defect slide create another slide that has what type it is (this is the correct answer). Set the transition times for 15 seconds for causes and 5 seconds for correct answers.

- Prepare the **cause** flash cards by writing on each card a one-sentence description of a cause using the key words from Section I. Create four to five causes per process problem type. Create a slide show with one cause per slide randomly ordered and after each cause slide create another slide that has what type it is (this is the correct answer). Set the transition times for 15 seconds for causes and 5 seconds for correct answers.
- Write instructions for participants on how to create a table on a sheet of paper or electronically that has five columns, one per process problem type, and two rows, the first labeled "Defects" and the second labeled "Causes."

Activities

1. Introduce the concept of reducing where to search by asking participants to describe their search to a solution of the problem: You go from your front door to your car but discover that you don't have the keys.
2. Introduce the five process problems by discussing the examples of people who already know how to solve the five problems.
3. Show and discuss the examples of defects with an explanation of the cause.
4. Give one example of each type of cause and ask what defect would result.
5. Have participants create the table on a sheet of paper. Run the **defect** slide show. Participants are to decide which type of process problem it is. If they are correct, they make a tick mark in the first row under the type of process problem.
6. Run the **cause** slide show. Participants are to decide which type of process problem it is. If they are correct, they make a tick mark in the second row under the type of process problem.
7. Discussion: Ask which type of problem they had the most difficulty in identifying. Ask why that was, and then review to address the obstacle. Provide more practice on the ones where they had the most difficulty.

8. Practice writing problem descriptions: Participants write an example of each type of problem by describing the cause and the defect. Participants review each other's results.

Note: You can repeat the second and third exercises to provide more practice after the discussion. You may adjust the times for the slides, decreasing as participants become more proficient. Fifteen seconds will seem a long time after a while.

Lesson 2: Requirements

Topic: Requirements

Learning Objective: After the lesson, participants will be able to state, analyze, and create the four components of a requirement and identify the type of process problem(s) that results when the requirement is not met.

Prerequisite: None

Class Length: Two (2) hours

Materials:

- An example of a requirement for each type of process problem with a word description
- Electronic requirement flash cards (20–25)
- Electronic problem type flash cards (20–25)
- Requirement Table
- Problem Type Table

Preparation

- Prepare the **requirement** flash cards by writing on each card an incomplete requirement that consists of a sentence with one of the four requirement components missing. Create four to five per process problem type using key words from Section I chapters. Write below each sentence the question: Which component is missing? Create a slide show with one requirement per slide randomly ordered and after each requirement slide create another slide that has the missing component (the correct answer). Set the transition times for 15 seconds for requirements and 5 seconds for correct answers.
- Prepare the **problem type** flash cards by using the requirement slide show and making two changes: (1) replace Which component is missing? with What type of process problem? and (2) replace each component answer slide with the problem type answer.

Set the transition times for 15 seconds for requirements and 5 seconds for correct answers. Reduce transition times for the slides after half the cards, if necessary.

- Write instructions for participants on how to create a Requirement Table on a sheet of paper or electronically that has four columns, one per component of a requirement.
- Write instructions for participants on how to create a Problem Type Table on a sheet of paper or electronically that has five columns, one per problem type.

Activities

1. Introduce the four components of a requirement by showing and discussing the examples.
2. Give one example of a requirement for each type of process problem.
3. Have participants create the requirement table on a sheet of paper. Run the requirement slide show. Participants are to decide which component is missing. If they are correct, they make a tick mark in the first row under the appropriate column.
4. Discussion: Ask which component they had the most difficulty in identifying. Ask why and then review to address the obstacle. Provide more practice on the ones they had difficulty with.
5. Have participants create the problem type table on a sheet of paper. Run the problem type slide show. Participants are to decide which type of process problem it is. If they are correct, they make a tick mark in the second row under the fifth column.
6. Discussion: Ask which type of problem they had the most difficulty in identifying. Ask why and then review to address the obstacle.
7. Practice writing requirements: Participants write an example of a complete requirement for each problem type. For example, they identify the type of process problem that would result if the requirement is not met. Participants review each other's results.

Lesson 3: Delays

Topic: Three-Step Procedure for Delays

Learning Objective: After the lesson, participants will be able to state and use the three-step procedure for solving defects due to delays on single units.

Prerequisite: Lessons 1–2. Each participant brings to the lesson one example of a delay that occurred in their own process.
Class Length: Each part can be done in approximately five hours, including several practice–feedback loops.
Materials:

- A written description of the three-step procedure
- One example of a documented application of the three-step procedure to a delay defect
- Documentation of delays in the registration process
- Each participant's example of a delay
- Three-Step Procedure Table: Column headings are "Example," "Problem Type," "Find Root Cause," "Address Root Cause"

Preparation

- Create a registration and confirmation process, but do not debug, practice, or mistake-proof it to ensure the process has errors and delays. Document the errors and delays. Have the students who suffered the problems either document the problem or provide them a copy of the documentation, e.g., acknowledge problems via electronic communication.

Activities

1. Immediately after introductions, ask if anyone had problems registering for the course. Select one participant who had a delay and have them describe the problem.
2. Introduce the principle of flow and the concept of jidoka.
3. Teach the three-step approach by applying it to the selected problem. In the first column of the table write a description of the example. Then, in the remaining three columns complete the three steps: identify type of problem, identify root cause, and solve root cause.
4. Select another delay problem. Have the participant describe the problem. Have participants work individually to find solutions using the three-step procedure.
5. Then have participants work in small groups to combine their ideas to create more powerful solutions.
6. Provide feedback. Identify where in the three-step procedure participants struggled. Discuss by asking other participants how they avoided or resolved the struggle areas. Summarize.
7. Repeat with a third delay problem.
8. Repeat with the delay problem participants brought to class.

9. Practice writing: Participants document their application of the three-step procedure by describing the type of problem, how to identify delay solutions, and how to find delay solutions. Participants review each other's results.

Lesson 4: Errors

Topic: Three-Step Procedure for Errors

Learning Objective: After the lesson, participants will be able to state and use the three-step procedure for solving defects due to errors on single units.

Prerequisite: Lessons 1–2. Each participant brings to the lesson one example of an error that occurred in their own process.

Class Length: Three (3) hours

Materials:

- A written description of the three-step procedure
- One example of a documented application of the three-step procedure to an error defect
- Documentation of errors in the registration process
- Each participant's one example of an error in their work processes
- Three-Step Procedure Table: columns headings are "Example," "Problem Type," "Find Root Cause," "Address Root Cause"

Preparation

- This lesson focuses solely on error-caused defects. The courses, for example, can be designed so people who register encounter both error and delay problems in the registration process. These defects then become the basis for teaching the concepts and providing practice.
- Create a registration and confirmation process, but do not debug, practice, or mistake-proof it to ensure the process has errors and delays. Document the errors and delays. Have the students who suffered the problems either document the problem or provide them a copy of the documentation, e.g., acknowledge problems via electronic communication.

Activities

1. Ask if anyone had problems registering for the course. Select one participant who had a delay and have them describe the problem.

2. Review the principle of flow and the concept of jidoka. Explain why single-unit process improvement is the basis for continuous improvement.
3. Teach the three-step approach by applying it to the selected problem: identify type of problem, identify root cause, and solve root cause.
4. Select another delay problem. Have the participant describe the problem. Have participants work individually to find solutions using the three-step procedure.
5. Then have participants work in small groups to combine their ideas to create more powerful solutions.
6. Provide feedback. Identify where in the three-step procedure participants struggled. Discuss by asking other participants how they avoided or resolved the struggle areas. Summarize.
7. Repeat with a third error problem.
8. Repeat with the error problem participants brought to class.
9. Practice writing: Participants document their application of the three-step procedure by describing the type of problem, how to identify error solutions, and how to find error solutions. Participants review each other's results.

Lesson 5: Suboptimality

Topic: Three-Step Procedure for Suboptimality

Learning Objective: After the lesson, participants will be able to state and use the three-step procedure for solving defects due to suboptimality on single units.

Prerequisite: Lessons 1–4. One example of suboptimality that occurred in their own process. A further restriction might be to those who have frequent occurrence of suboptimality in their processes, e.g., R&D, manufacturing.

Class Length: One day

Materials:

- A written description of the three-step procedure
- One example of a documented application of the three-step procedure to suboptimality defect
- One-factor toast problem
- Two-factor toast problem

- Each participant's example of suboptimality in their work processes
- Three-Step Procedure Table: columns headings are "Example," "Problem Type," "Find Root Cause," "Address Root Cause"

Preparation

- Write on paper or in an electronic file the description of the one-factor toast problem: Your toast was too dark after toasting in a toaster oven. The toaster oven has a toast option, but there is only one control knob that ranges from Light when horizontally pointing to the left and Dark when horizontally pointing to the right. The knob had been set to Dark.
- Write on paper or in an electronic file the description of the two-factor toast problem: Your toast was too dark after toasting in a toaster oven. The toaster oven has a toast option and there are two ways to control the darkness. There is a control knob that ranges from Light when horizontally pointing to the left and Dark when horizontally pointing to the right. The knob had been set to Dark. There is a power setting that ranges from 1 (low) to 9 (high). It had been set at 9.

Activities

1. Review the principle of flow and the concept of jidoka. Explain why single-unit process improvement is the basis for continuous improvement.
2. Review examples of suboptimality process problems with electronic flash cards.
3. Review the three-step approach by applying it to an error problem: identify type of problem, identify root cause, and solve root cause.
4. Teach basic concepts of design of experiment.
5. Review the three-step approach by applying it to a one-factor suboptimality problem: identify type of problem, identify root cause, and solve root cause.
6. Have participants individually apply the three-step procedure to a one-factor toast suboptimality problem. Then have participants work in small groups to review their result for each step and combine their ideas to create more powerful solutions.
7. Provide feedback. Identify where in the three-step procedure participants struggled. Discuss by asking other participants how they avoided or resolved the struggle areas. Summarize.
8. Review two-factor design of experiments. Repeat application of three-step procedure with the two-factor toast suboptimality problem.

Then have participants work in small groups to review their result for each step and combine their ideas to create more powerful solutions.

9. Provide feedback. Identify where in the three-step procedure participants struggled. Discuss by asking other participants how they avoided or resolved the struggle areas. Summarize.
10. Review three-factor design of experiments. Repeat application of three-step procedure with the two-factor toast suboptimality problem, but ask them to identify another factor that would influence how dark the toast would be (e.g., time). Then have participants work in small groups to review their result for each step and combine their ideas to create more powerful solutions.
11. Provide feedback. Identify where in the three-step procedure participants struggled. Discuss by asking other participants how they avoided or resolved the struggle areas. Summarize.
12. Repeat with the suboptimality problem that participants brought to class.
13. Practice writing: Participants document their application of the three-step procedure by describing the type of problem, how to identify suboptimality solutions, and how to find suboptimality solutions. Participants review each other's results.

Lesson 6: Unpredictability

Topic: Three-Step Procedure for Unpredictability

Learning Objective: After the lesson, participants will be able to state and use the three-step procedure for solving defects due to unpredictability on single units.

Prerequisite: Lessons 1–5. One example of unpredictability that occurred in their own process. A further restriction might be to those who have frequent occurrence of unpredictability in their processes, e.g., manufacturing demand, sales.

Class Length: One day

Materials:

- A written description of the three-step procedure
- One example of a documented application of the three-step procedure to unpredictability defect
- Two unpredictability processes' measures: participants' arrival time to class, ending time of class

- Each participant's example of unpredictability in their work processes
- Three-Step Procedure Table: columns headings are "Example," "Problem Type," "Find Root Cause," "Address Root Cause"

Preparation

- Arrival: Instructions for participants include the question: Why do participants not arrive between 15 minutes before class and the start time of class?
- Ending: Instructions for participants include the question: Why does the class not end on the scheduled time or no earlier than 15 minutes before scheduled time?

Activities

1. Review the principle of flow and the concept of jidoka. Explain why single-unit process improvement is the basis for continuous improvement.
2. Review examples of unpredictability process problems with electronic flash cards.
3. Review the three-step approach by applying it to an error problem: identify type of problem, identify root cause, and solve root cause.
4. Teach basic concepts of correlation and regression.
5. Review the three-step approach by applying it to a one-factor unpredictability problem: identify type of problem, identify root cause, and solve root cause.
6. Have participants individually apply the three-step procedure to the simulation unpredictability problem. Then have participants work in small groups to review their result for each step and combine their ideas to create more powerful solutions.
7. Provide feedback. Identify where in the three-step procedure participants struggled. Discuss by asking other participants how they avoided or resolved the struggle areas. Summarize.
8. Review two-factor correlation analysis. Repeat application of three-step procedure with the two-factor unpredictability problem. Then have participants work in small groups to review their result for each step and combine their ideas to create more powerful solutions.
9. Provide feedback. Identify where in the three-step procedure participants struggled. Discuss by asking other participants how they avoided or resolved the struggle areas. Summarize.
10. Review three-factor correlation analysis. Repeat application of three-step procedure with the three-factor unpredictability problem.

Then have participants work in small groups to review their result for each step and combine their ideas to create more powerful solutions.

11. Provide feedback. Identify where in the three-step procedure participants struggled. Discuss by asking other participants how they avoided or resolved the struggle areas. Summarize.
12. Repeat with the unpredictability problem participants brought to class.
13. Practice writing: Participants document their application of the three-step procedure by describing the type of problem, how to identify unpredictability solutions, and how to find unpredictability solutions. Participants review each other's results.

Lesson 7: Personal Reason

Topic: Three-Step Procedure for Personal Reasons

Learning Objective: After the lesson, participants will be able to state and use the three-step procedure for solving defects due to personal reasons on single units.

Prerequisite: Lessons 1–6. One example of personal reason that occurred in their own process.

Class Length: Two (2) hours

Materials:

- A written description of the three-step procedure
- One example of a documented application of the three-step procedure to a personal reason defect
- One personal reason case whose reason is based on another type of defect
- One personal reason case that is not based on another type of defect
- Each participant's example personal reason defect in their work processes
- Three-Step Procedure Table: columns headings are "Example," "Problem Type," "Find Root Cause," "Address Root Cause"

Preparation

- Other defect type: Write instructions for participants. Form pairs where one person is the Customer and the other is the Vendor. The Customer's instructions state that the customer bought an item, but did not pay the invoice because the item was defective. The Vendor's instructions state that the vendor views the nonpayment as a defect in the invoicing process.

- Personal Reason defect type: Write instructions for participants. Form pairs where one person is the Customer and the other is the Vendor. The Customer's instructions state that the customer decided not to renew membership. The customer gets to decide the personal reason for not renewing. The Vendor's instructions state that the vendor views the nonrenewal as a defect in the sales process.

Activities

1. Review the three-step procedure by asking participants to describe the steps and deliverables.
2. Review the personal reason example process.
3. Apply the three-step procedure to the first personal reason case. Document deliverables after each step. Review, discuss, and provide feedback after each step.
4. Apply the three-step procedure to the second personal reason case. Document deliverables after each step. Review, discuss, and provide feedback after each step.
5. Apply the three-step procedure to the participants' own processes with personal reason defects. Document deliverables after each step. Review, discuss, and provide feedback after each step.
6. Provide feedback. Identify where in the three-step procedure participants struggled. Discuss by asking other participants how they avoided or resolved the struggle areas. Summarize.
7. Practice writing: Participants document their application of the three-step procedure by describing the type of problem, how to identify error solutions, and how to find error solutions. Participants review each other's results.

Chapter 12

Critical Thinking Lessons

Lessons 8 through 14 of this chapter follow the material in Section II on critical thinking. Make sure that Lesson 2 on requirements is taught before these lessons. Use the book as a reference for class participants, including developing materials from it, e.g., chapter introductions, examples, tables, figures, etc.

Lesson 8: Critical Thinking

Topic: Critical Thinking

Learning Objective: After the lesson, participants will be able to describe the critical thinking questions, the information needed to answer the questions, and cite the sources for the information. Participants also will be able to explain the basis for critical thinking and the connection to the three-step procedure used for solving process problems on single units.

Prerequisite: Lessons 1–7.

Class Length: Two (2) hours

Materials:

- A written description of the four critical thinking questions including purpose and information needed to answer each question
- One example of a requirement for each type of process problem
- One example of baseline performance, desired performance, and consequence statements for each type of process problem
- One example each of prioritization by criticality, consequence, defect rate
- One example of a control for each problem type
- Critical thinking table of questions, information, and sources

Preparation: Create examples, descriptions, and table.
Activities

1. Review the principle of flow and introduce the concept of *heijunka*. Explain chronic problems and their connection to multiple units.
2. Introduce the concept of problem solving as information-based decision making. Explain how critical thinking exemplifies this concept: Critical thinking consists of questions about how you know certain things and answers that are based on information.
3. Show the connection between critical thinking and the three-step procedure the participants already know using the critical thinking table of questions, information, and sources.
4. Introduce first critical thinking question: "How do I know I have a process performance problem?" and answer: "Because I know I have a performance gap that is worth closing."
   a. Introduce the concept of requirements. Provide examples and practice with flash cards on requirements.
   b. Introduce the concept of process performance, current process performance, and desired process performance. Provide examples and practice with flash cards on current and desired process performance statements.
5. Introduce second critical thinking question: "How do I know the root cause?" and answer: "Because I know which causal relationship improves process performance with respect to the requirement."
   a. Show that they already know how to do this—the second step of the three-step procedure for single units.
   b. Review identifying problem type for requirements.
   c. Review how to find root causes by problem type.
   d. Introduce data collection and random sampling. Provide practice on data collection.
6. Introduce third critical thinking question: "How do I know my solution works?" and answer: "Because I know that I can sustain closing the gap with the proposed process change."
   a. Show that they already know how to find solutions—the third step of the three-step procedure for single units.
   b. Review identifying solutions by problem type.
   c. Introduce controls and their purpose. Provide examples of controls by problem type and practice creating controls.

7. Introduce fourth critical thinking question: "How do I know when I can improve again?" and answer: "Because I know there is an unplanned improved performance for which I can find the root cause."
   a. Show that they already know how to do this—the three-step procedure for single units and second and third critical thinking questions.
   b. Review controls and their second purpose to signal when an unplanned improvement occurred. Provide examples of controls by problem type signaling improvements and practice creating controls.

Lesson 9: Delays

Topic: Critical Thinking for Delays

Learning Objective: After the lesson, participants will be able to state and use critical thinking questions for solving defects due to delays on multiple units

Prerequisite: Lessons 1–4 and 8. One example of delays that occurred in their own process.

Class Length: Five (5) hours

Materials:

- A written description of the four critical thinking questions including purpose and information needed to answer each question
- One example of a documented application of critical thinking to a delay defect
- Documentation of delays in the registration process
- Each participant's example of delays
- One example of baseline performance, desired performance, and consequence statements for each type of process problem
- One example of a control for delays
- Critical thinking table

Preparation: Select examples from Lesson 3 and simulations.

Delay Simulations:

Materials: Lined and plain paper, pens/markers, Post-Its® (3 x 5 in., six different colors), tape, scissors.

Process: Take orders from customers, invoice customers, collect payment.

Setup: Two or more companies compete to service multiple customers; customers order quantities of a product (e.g., bottled water of different amounts and different quantities; product is not actually delivered or produced); price per amount is fixed, but let the companies compete on price if they ask; customers can refuse to pay if invoices are incorrect and they should attempt to order frequently enough to stress the processes (e.g., every two to three minutes). This is a cash system so customers have to write "checks."

Business Performance: Companies are competing for market share and revenue, so determine market share by orders collected; revenue is cash collected not invoices sent.

Simulation: Assign people to or ask for volunteers for roles (customer, sales, and accounting). Do NOT provide instructions or limitations on what they can do except that they have to do the simulation. Allow at least one hour for participants to design processes for their roles of getting orders, submitting invoices, and collecting payments. When they are ready, run the simulation for 30 minutes. Evaluate performance on how long it takes to process. Apply critical thinking questions to improve. Allow them time to reset. Run simulation for 30 minutes with improved processes. Compare before versus after results. Document lessons learned including what would they do differently in designing new processes.

Variations: (1) Add suppliers (order raw materials, invoice for them, and collect) who give the customer a delivery notice rather than actual product. Customer can refuse notice or to pay invoice if delivery notice is not correct in what was ordered; suppliers charge a fixed price per bottle and amount of water (e.g., per ounce); also evaluate business performance by net income of cash on hand and paid using cash basis for payment to suppliers; (2) if you have participants from the same function or department, use actual processes.

Activities

1. Review the critical thinking procedure by having participants teach the questions, the information, and the sources.
2. Apply critical thinking to delays in the registration process. Document deliverables after each step. Review, discuss, and provide feedback after each step.

3. Run simulation. Apply critical thinking to delays in the simulation. Document deliverables after each step. Review, discuss, and provide feedback after each step.
4. Apply critical thinking to delays in participants' own processes. Document deliverables after each step. Review, discuss, and provide feedback after each step.

Lesson 10: Errors

Topic: Critical Thinking for Errors

Learning Objective: After the lesson, participants will be able to state and use critical thinking questions for solving defects due to errors on multiple units.

Prerequisite: Lessons 1–4 and 8–9. One example of errors that occurred in their own process.

Class Length: Five (5) hours

Materials:

- A written description of critical thinking questions including purpose and deliverables
- One example of a documented application of critical thinking questions to an error defect
- Documentation of errors in the registration process
- Each participant's example of errors
- Two or more simulations
- One example of baseline performance, desired performance, and consequence statements for each type of process problem
- One example of a control for errors
- Critical thinking table

Preparation: Select examples from Lesson 4 and simulations.

Error Simulation:

Materials: Lined and plain paper, pens/markers, Post-Its (3 x 5 in., six different colors), tape, scissors.

Process: Take orders from customers, invoice customers, collect payment.

Setup: Two or more companies compete for servicing multiple customers; customers order quantities of a product (e.g., bottled water of different amounts and different quantities—product is not actually delivered or produced); price per amount is fixed, but let the companies compete on price if they ask; customers can refuse to pay if invoices are incorrect and they should attempt to order frequently enough to stress the processes (e.g., every two to three minutes). This is a cash system so customers have to write "checks."

Business Performance: Companies are competing for market share and revenue, so determine market share by orders collected; revenue is cash collected not invoices sent.

Simulation: Assign people to or ask for volunteers for roles (customer, sales, and accounting). Do NOT provide instructions or limitations on what they can do except that they have to do the simulation. Allow at least one hour for participants to design processes for their roles of getting orders, submitting invoices, and collecting payments. When they are ready, run the simulation for 30 minutes. Evaluate performance on how long it takes to process. Apply critical thinking questions to improve. Allow them time to reset. Run simulation for 30 minutes with improved processes. Compare before versus after results. Document lessons learned including what would they do differently in designing new processes.

Variations: (1) Add suppliers (order raw materials, invoice for them, and collect) who give the customer a delivery notice rather than actual product. Customer can refuse notice or to pay invoice if delivery notice is not correct in what was ordered; suppliers charge a fixed price per bottle and amount of water (e.g., per ounce); also evaluate business performance by net income of cash on hand and paid using cash basis for payment to suppliers; (2) if you have participants from the same function or department, use actual processes; Create the simulation material.

Activities

1. Review the critical thinking procedure by having participants teach the questions, the information, and the sources.
2. Apply critical thinking to errors in the registration process. Document deliverables after each step. Review, discuss, and provide feedback after each step.
3. Run simulation. Apply critical thinking to errors in the simulation. Document deliverables after each step. Review, discuss, and provide feedback after each step.

4. Apply critical thinking to errors in participants' own processes. Document deliverables after each step. Review, discuss, and provide feedback after each step.

Lesson 11: Suboptimality

Topic: Critical Thinking for Suboptimality

Learning Objective: After the lesson, participants will be able to state and use critical thinking question for solving defects due to suboptimality on multiple units.

Prerequisite: Lessons 1–5 and 8–10. One example of suboptimality that occurred in their own process.

Class Length: Five (5) hours

Materials:

- A written description of critical thinking question including purpose and deliverables
- One example of a documented application of critical thinking questions to a suboptimality defect
- Each participant's example of suboptimality
- One example of baseline performance, desired performance, and consequence statements for each type of process problem
- One example of a control for suboptimality
- Two suboptimality simulations
- Critical thinking table

Preparation: Select examples from Lesson 5 and simulations.

Suboptimal Setting Simulations:

Materials: Two paper types of different weight, pens/markers, scissors, tape, paper clips, computer stopwatch (search Web for free downloads).

Process: Create helicopter with maximum flight time exceeding (LSL) 2.4 seconds by finding optimal design.

Setup: Assign people to teams of three to five people per team. Describe to the teams the basic design (Figure 12.1). An 8.5 x 11 in. paper is cut in half lengthwise (portrait) to make two helicopters. Each helicopter has three sections: wings (cut along dotted line to separate and then fold in opposite directions), body, and tail. Each of these can vary in length and width. In addition to these six factors, you can add other factors, e.g., paper clip the tail for weight, tape at the wing junction to hold it steady.

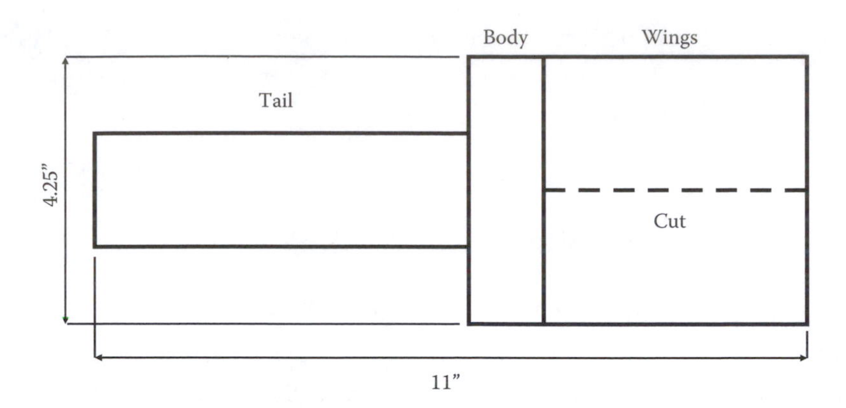

Figure 12.1 Helicopter.

Simulation: (1) Teams select the settings and the combinations of the following seven factors: wing length, wing width, body length, body width, tape on wings, tape on body, and paper clip on body. Each team creates 15 to 20 helicopter designs. Each team determines from what altitude they will drop the helicopters (recommend at least eight feet). Teams drop their helicopters, recording the time to land and repeat for at least six drops (flights) per helicopter. (2) Teams enter their data in their statistical software: the settings for each helicopter and the times for each flight. They use statistical hypothesis testing to determine which factors are correlated with average flight time, variation in flight time, and percent good (select a LSL (lower specification limit) = 2.4 for eight-foot drops). (3) Teams check whether any of the correlated factors are causes of longer flight time using a design of experiment (DOE) with two levels per factor. Once they have the critical factors, if they need to determine optimal settings for the quantitative factors, they use a response surface DOE.

Variations: You can shorten the course and focus on DOE by providing the historical data; if you have participants from the same function or department, use actual processes; for people in R&D or manufacturing, redevelop existing products and lines to see what would have been done differently and what new insight is there to these products and lines that can be used to improve them (preventive action) or to incorporate into corrective action procedures; search the Web for DOE case studies, e.g., http://courses.ncssm.edu/math/Stat_Inst/PDFS/PaperTow.pdf *or* http://www.statease.com/pubs/doe-self.pdf.

Suboptimal Procedure Simulation:

Materials: Lined and plain paper, pens/markers, Post-Its (3 x 5 in., six different colors), tape, scissors.

Process: Take orders from customers, invoice customers, collect check payment.

Setup: Group people into sets of three.

Simulation: Within each group of three, assign roles of customer, sales rep, and accountant. Customers will create orders and pay invoices; sales will take orders from customer back to accounting, take invoices from accounting back to customer, and checks from customer back to account; accounting will create invoices based on orders, receive checks for orders, and keep track of payments for a total amount collected at end of simulation. Do NOT provide instructions or limitations on what they can do except that they have to do the simulation in the room. Allow 30 to 60 minutes for participants to design processes for their roles of getting orders, submitting invoices, and collecting payments. Tell them that each set of three is a team competing against the other teams on speed and accuracy. When they are ready, run the simulation for 30 minutes. Evaluate performance on how long it takes to process and how many errors occur. Apply critical thinking questions to improve; analyze to determine who did best overall or by role and then standardize to best procedure. Allow them time to reset. Run simulation for 30 minutes with improved standard operating procedure (SOP). Compare before versus after results. Document lessons learned including what would they do differently in designing new processes.

Variations: If you have participants from the same function or department, use actual processes; for manufacturing simulation, form pairs for roles of designer (creates instructions to build paper airplane or helicopter or use materials, e.g., Legos®) and production (creates product based on instructions) with production creating 5 to 10 copies of product using the same instructions competing on speed and time.

Activities

1. Review the critical thinking procedure by having participants teach the questions, the information, and the sources.
2. Apply critical thinking to the two-factor toast (Lesson 3) process. Document deliverables after each step. Review, discuss, and provide feedback after each step.

3. Apply critical thinking to the suboptimality of the simulation. Document deliverables after each step. Review, discuss, and provide feedback after each step.
4. Apply critical thinking to suboptimality of the participants' own processes. Document deliverables after each step. Review, discuss, and provide feedback after each step.

Lesson 12: Unpredictability

Topic: Critical Thinking for Unpredictability

Learning Objective: After the lesson, participants will be able to state and use critical thinking questions for solving defects due to unpredictability on multiple units.

Prerequisite: Lessons 1–6 and 8–11. One example of unpredictability that occurred in their own process.

Class Length: One (1) day

Materials:

- A written description of critical thinking questions including purpose and deliverables
- One example of a documented application of critical thinking questions to a unpredictability defect
- Each participant's example of unpredictability
- One example of baseline performance, desired performance, and consequence statements for each type of process problem
- One example of a control for unpredictability
- Two unpredictability simulations
- Critical thinking table

Preparation: Select examples from Lesson 6 and simulations.

Unpredictability Simulations:

Materials: Laptop with Excel®

Process: As a travel advisor, you provide information to airline travelers on how to reduce the risk of flight delays. You do not have any control over the causes (e.g., weather and mechanical problems), so you can only use past performance to predict future performance. Travelers want to know which airlines and airports will maximize their chances of on-time arrivals.

Setup: Assign people to teams of three to five people per team. Each team will select five airports, four airlines, and one to two months of recent data from the Bureau of Transportation Statistics (BTS), online at: http://www.bts.gov/xml/ontimesummarystatistics/src/dstat/OntimeSummaryArrivals.xml. The data can be downloaded into Excel before loaded into the statistics software. Provide them with a margin of error, e.g., ± 10%, and specifications for on-time arrival = Actual Arrival Time – Scheduled Arrival Time, e.g., USL (upper specification limit) = 5 minutes.

Simulation: Instructions for each team: (1) Randomly select 20 to 25% of the data and set aside. Use the remaining 75 to 80% of the data to develop a prediction model for Actual Arrival Time – Scheduled Arrival Time. (2) Validate your model using the data you set aside. Does it predict within the margin of error at the desired percent good? If so, determine what airline and airport will maximize the percent of on-time. If not, develop another model.

Variations: You can shorten the course and focus on modeling by providing the data; if you have participants from the same function or department, use actual processes; for practice with qualitative predictions, use data on flight cancellations (http://www.bts.gov/xml/ontimesummarystatistics/src/dstat/OntimeSummaryCancellation.xml) to determine which airport and airline minimizes chance of cancellation; extend the number of factors by considering time of day or week and/or flight number; use financials ratios to predict stock performance; order data from NHTSA (National Highway Traffic Safety Administration) on car accidents and develop simulation to predict accidents (they use SAS®, so data may need to be converted). Also, search the Web for artificial intelligence in prediction models.

Activities

1. Review the critical thinking procedure by having participants teach the questions, the information, and the sources.
2. Apply critical thinking to the unpredictability of the registration process. Document deliverables after each step. Review, discuss, and provide feedback after each step.
3. Run simulation. Apply critical thinking to the unpredictability of the simulation. Document deliverables after each step. Review, discuss, and provide feedback after each step.
4. Apply critical thinking to the unpredictability of the participants' own processes. Document deliverables after each step. Review, discuss, and provide feedback after each step.

Lesson 13: Personal Reason

Topic: Critical Thinking for Personal Reason

Learning Objective: After the lesson, participants will be able to state and use critical thinking questions for solving defects due to personal reasons on multiple units.

Prerequisite: Lessons 1–12. One example of personal reason that occurred in their own process.

Class Length: Four (4) hours

Materials:

- A written description of critical thinking questions including purpose and deliverables
- One example of a documented application of critical thinking questions to a personal reason defect
- Each participant's example of a personal reason
- One example of baseline performance, desired performance, and consequence statements for each type of process problem
- One example of a control for personal reason
- Two personal reason simulations: (1) reason is based on another type of defect and (2) reason is not based on another type of defect
- Critical thinking table

Preparation: Select examples from Lessons 1–13 and simulations.

Personal Reason Simulations:

Materials: Paper, pens.

Process: Employment.

Setup: Form pairs; after first round, form new pairs.

Simulation: In each pair, one person is the Asker and the other person is the Responder. The Asker asks the Responder why they would leave the company. Asker records the Responder's reason(s). Switch roles, repeat. Form new pairs, and repeat role-playing. After discussion, determine whether repeating is of value. After data are gathered, form small groups. Create a list of the reasons and evaluate them: (1) Does the reason identify a defect? If so, what defect and what type of problem is it? (2) If not, determine whether you would be willing and able to address the reason. If so, state how.

Variations: Have participants select their own customer situation and provide the reason (can be done as pairs or in small groups).

Activities

1. Review critical thinking questions by having participants teach each step.
2. Review the personal reason example process.
3. Run simulation. Apply critical thinking questions to the first personal reason simulation. Document deliverables after each step. Review, discuss, and provide feedback after each step.
4. Apply critical thinking questions to the second personal reason simulation. Document deliverables after each step. Review, discuss, and provide feedback after each step.
5. Apply critical thinking questions to participants' own processes where personal reasons occur. Document deliverables after each step. Review, discuss, and provide feedback after each step.

Lesson 14: Design

Topic: Process Design

Learning Objective: After the lesson, participants will be able to state and use critical thinking questions for designing processes for multiple units with reduced probability of defects for all five types of problems.

Prerequisite: Lessons 1–13. One example of process design of their own processes.

Class Length: One (1) day

Materials:

- A written description of critical thinking questions including purpose and deliverables
- One example of a documented application of critical thinking questions to process design
- Each participant's example of a process design
- One example of baseline performance, desired performance, and consequence statements for each type of process problem
- One example of a control for each type of process problem
- Two process design simulations
- Critical thinking table

Preparation: Select examples from Lessons 1–13 and simulations.

Process Design Simulations

Materials: Paper, markers, Post-Its®.

Processes: Registration, arrival to class, breaks, distribution of materials.

Activities

1. Review the three-step procedure by asking participants to describe the steps and deliverables.
2. Review the principle of flow and introduce the concept of heijunka. Explain how chronic problems are related to multiple units and the need to go beyond single-unit process improvement.
3. Review critical thinking questions by having participants teach each step.
4. Review the example process design.
5. Run simulation. Apply critical thinking questions to the first process design simulation. Document deliverables after each step. Review, discuss, and provide feedback after each step.
6. Run second simulation. Apply critical thinking questions to the second process design simulation. Document deliverables after each step. Review, discuss, and provide feedback after each step.
7. Apply critical thinking questions to participants' own processes they are to design. Document deliverables after each step. Review, discuss, and provide feedback after each step.

Detail Tasks for Activities 4–7:

a. Identify Requirements. Talk to customers to get their requirements. This includes determining the customer specifications. Repeat with the business to get its requirements with specifications.
b. Determine VA actions from the requirements. Identify only those actions that add value to the customer (CVA). Repeat for BVA (business value-added) actions.
c. Determine which value-added (VA) actions can have errors; develop mistake-proofs.
d. Determine which VA actions need optimization; use DOE on critical factors.
e. Determine which VA actions depend on predictability; use correlation analysis to identify predictors and develop model; develop procedure for using model and continual confirmation of its predictability.
f. Determine which VA actions could have defects for personal reasons; develop procedure for asking and determining whether the reason is due to another type of defect or is a real personal reason; develop criteria for determining which real personal reasons you are willing and able to address.
g. Attempt to design process with only VA actions.

h. First, include only CVA actions.
i. Second, integrate the BVA actions.
j. If you cannot design process with only VA actions, add NVA actions.
k. Assess process time and use what you learned about delay-caused defects to reduce total time.

Chapter 13

Mentoring

Mentor the person on the process. This means do not engage as a subject matter expert. This isn't about you or your knowledge, it's about teaching and facilitating. Use the critical thinking questions to guide and facilitate your mentee.

As a mentor, asking leading questions and asking for explanations helps mentees develop their critical thinking through deeper understanding. Ask first, rather than provide answers.

Preparation

Keep the critical thinking table (Table sec. II.1) with you. Use it to focus discussions and interactions. At all times, try to be physically on the same side of a table as your mentee to embrace the idea that you are tackling the problem together. Use the pronoun "we" rather than "you" in discussing progress and next steps.

Approach

Your task is to guide the mentees through the four critical thinking questions. Help them by asking them questions. You can start each meeting or interaction with these questions:

a. Which critical thinking question are you trying to answer?
b. What information do you need to answer the question?

c. Where is the information?
d. How will you get that information?
e. How will you know when you have that information?
f. Can you answer the critical thinking question now?

(a) *Which critical thinking question are you trying to answer?* If they do not know or cannot remember the exact wording, use the critical thinking table to remind them. Repetition and visual reminders helps a person remember.

Help them understand the importance of the question by asking mentees why they need to answer this question and why now.

(b) *What information do you need to answer the question?* If they do not know or cannot remember all the information they need, use the critical thinking table (Table sec. II.1) to remind them. Repetition and visual reminders helps a person remember. Ask them which information they need first and why.

(c) *Where is the information?* If they do not know or cannot remember what sources can provide them the information they need, use the critical thinking table (Table sec. II.1) to remind them. Repetition and visual reminders helps a person remember. Table sec. II.1 lists sources generically. They will need to know more specifically where the information is. Ask them where it is specifically and why they think it is there.

(d) *How will you get that information?* If they do not know or cannot remember how, ask them for ideas. Ask them to be specific in explaining how. If you can think of any potential difficulty because they are not specific enough, ask them how they will address the obstacle. Ask them why the actions they suggest will get them the information.

If they need tools, ask what tool might or should they use. Lead them to an answer by asking questions. For example, you can ask these questions:

- What is the purpose of the tool?
- Will that purpose produce the information you need?
- What assumptions are required to use the tool to produce the information you need?
- Are the assumptions supported?

(e) *How will you know when you have that information?* The answer is: "Because they will be able to answer the critical thinking question." If they do not know or cannot remember what sources can provide them the information they need, use the critical thinking table to remind them. Repetition and visual reminders helps a person remember.

(f) *What is the answer to the critical thinking question?* The answer will confirm (or not) that they have used critical thinking.

Summary

The key to successful mentoring is to guide mentees through the process, procedure, or methodology they are using. In solving process problems, the four critical thinking questions define the method. Guide by asking them what question they want to answer, what information they need to answer the question, and how will they get that information.

Recap

It is easier to solve a problem in your house if you know the type of the problem (plumbing, electricity, or infestation problem), and then you can solve it faster.

Similarly, with process problems, knowing the type of problem makes solving it easier.

Problems are simply gaps between what you have and what you desire. There are two levels of process problems. A single output can have a defect that you don't want or you can have chronic defects on multiple units that you don't want. Both levels of process problems result in a gap between your current performance and your desired performance.

A defect is not meeting a requirement. A requirement has four components:

1. Output unit (product or service unit)
2. Characteristic of interest (e.g., related to time, function, fit, form)
3. Measure (the choices for the characteristic, e.g., blue or red for color or 1 cubic meter for volume)
4. Specifications (the criteria that determines if it is defective or not)

There are five types of defects. Each type is defined by its cause:

1. Error-caused defects
2. Delay-caused defects
3. Suboptimality-caused defects
4. Unpredictability-caused defects
5. Personal Reason-caused defects

Table Recap.1 Key Activities for Solving Process Problems Fast and Effectively

Type of Defect: Cause	*How To Identify: Key Words*	*How to Determine Root Cause*	*How to Address Root Cause*
Error	Accuracy, mistake, error, wrong	Retrace	Mistake-proof
Delay	On-time delivery, too long (e.g., lead time), too slow (e.g., pace, changeover, setup), backlog	Identify the type of delay: Flow stopping Rework Nonvalue-added actions Slow value-added actions	Address the type of delay: Remove barriers to continuous flow Reduce defects (see other types of defects) Remove NVA actions Replace VA actions with faster ones
Suboptimality	Scrap/waste, quality, rework, capacity, uptime low/downtime high, chronic or sporadic or intermittent failure, can't make product, unplanned maintenance, broken (process, step, machine, component, part)	DOE: Use scientific laws and theories, historical data, and data-based experiences to select potential factors; then use DOE to identify and confirm root causes and their optimal levels	Set critical factors to optimal levels
Unpredictability	Customer demand (variable, high, low), too many/few (products/types, resources including people), schedule (variation), inventory (off, low, excessive)	Correlation analysis: Use subject matter expertise and experience to select potential predictors; then use correlation analysis to confirm predictors and build model	Use current levels of predictors in prediction model

Table Recap.1 (*Continued*) Key Activities for Solving Process Problems Fast and Effectively

Type of Defect: Cause	*How To Identify: Key Words*	*How to Determine Root Cause*	*How to Address Root Cause*
Personal Reason	Accounts receivable (high), employee turnover (high), customer loyalty (low), policy (violation), compliance	Ask	If the reason is due to one of the other causes, address that cause If not, then determine whether you are willing and able to address the reason

Improving a process means improving its performance. Process performance is defined by the percent of good things (output units) it produces. So, reducing the frequency of defects improves process performance.

For single-unit defects, you can reduce defects more effectively and quicker by following a three-step procedure:

1. Identify the type of defect: Use the key words in the chapters of Section I to help you.
2. Determine the root cause: Use the type of defect to shorten your search for root causes.
3. Address the root cause: Use the type of cause to quickly identify solutions.

Table Recap.1 summarizes the key activities for solving process problems fast and effectively.

When solving chronic defects, use critical thinking to answer four questions about each type of defect:

1. How do I know I have a process performance problem?
2. How do I know the root cause?
3. How do I know the proposed solution works?
4. How do I know when I can improve again?

You can also use critical thinking to design new processes with lower defect rates for all five types of problems.

Now it's up to you. Go and improve processes. Get rid of those defects.

Index

D

L

M

N

Q

R

T

U

About the Author

Kicab Castaneda-Mendez is a business and process improvement professional with 30 years experience as an internal and external consultant to manufacturing, service, healthcare, government, and nonprofit organizations from 21 industries on five continents. He has authored two books and more than 35 articles, spoken at more than 25 conferences, and given more than 50 conference workshops on theory, tools, and applications on achieving performance excellence. He has taught several thousand people at every level from hourly to CEOs and has helped improve processes in the supply chain (marketing, vendors, R&D, engineering, QA/RA, manufacturing, sales), support functions (logistics, HR, IT, legal, finance), and management. He is a three-year Baldrige examiner and five-year Connecticut Award for Excellence senior examiner and trainer. Castaneda-Mendez is a GE-certified Master Black Belt, MBB/BB/GB trainer, and facilitator. He has two master's degrees (statistics and mathematics), a triple major bachelor's degree (mathematics, philosophy, and psychology), and a permanent Secondary teaching Certificate (mathematics) all from The University of Michigan.